EXPLORING LIGHT Investigations for Optics Education

SECOND EDITION

EXPLORING LIGHT
Investigations for Optics Education

SECOND EDITION

JUDY DONNELLY
NANCY MAGNANI

SPIE PRESS
Bellingham, Washington USA

Library of Congress Control Number: 2024944939

Published by
SPIE
P.O. Box 10
Bellingham, Washington 98227-0010 USA
Phone: +1 360.676.3290
Fax: +1 360.647.1445
Email: books@spie.org
Web: www.spie.org

First edition, *Explorations in Optics* © 2005 Judith Donnelly published by New England Board of Higher Education.

Printed in the United States of America.
First printing 2025.
For updates to this book, visit http://spie.org and type "PM383" in the search field.

SPIE.

Contents

https://t.ly/6Lio8

indicates that there is a video of this experiment on the PBL Projects YouTube™ channel,[1] which you can find using the QR code and link above.

Acknowledgments

September 2024

The first edition of this book, *Explorations in Optics*, was published by the New England Board of Higher Education (NEBHE) as a result of the **PHOTON Projects** and **Problem-Based Learning (PBL) Projects**, which were funded in part by grants from the National Science Foundation Division of Undergraduate Education (DUE) Advanced Technological Education (ATE) program. We are grateful to NEBHE and especially to now retired Senior Director of Programs Fenna Hanes for supporting this work and continuing to host our website, pblprojects.org.

We also thank SPIE, the international society for optics and photonics, especially former Manager of Education Services Kathleen Robinson (who gave the *Dumpster Optics* lessons their name) for supporting our workshops for many years at the SPIE Optics and Photonics conference and at the Education and Training in Optics and Photonics (ETOP) conference.[2] Thanks also to the Optica (formerly OSA) Foundation for the grant that allowed us to hone these experiments and demonstrations by practicing with more than 10 years of workshops for hundreds of young students each year. Optica also provided funding to record the Optics Exploration videos with students from Eastern Connecticut State University. Thanks also to Groot Gregory, product manager at Synopsys and SPIE Fellow, for outreach inspiration, including the bedazzled kaleidoscope and graded index gelatin. Finally, we are grateful to Dr. Perla Marlene Viera González and Dr. Guillermo Ezequiel Sanchez-Guerrero of Universidad Autónoma de Nuevo León, Mexico, for her enthusiastic outreach and for translating our work into Spanish.

www.pblprojects.org

The PHOTON Projects [Fiber Optic Technology Education Project (FOTEP), Division of Undergraduate Education (DUE) #9553762; PHOTON, DUE #0053284, PHOTON2 DUE #0302528; PHOTON PBL, DUE #0603143] and the PBL Projects (STEM PBL, DUE #0903051; Advanced Manufacturing PBL, DUE #1204941; Advanced Photonics Manufacturing PBL, DUE #1801115) were funded by the National Science Foundation's Advanced Technological Education program (NSF ATE). Any opinions, findings, and conclusions or recommendations expressed in this material are those of the authors and do not necessarily reflect the views of the National Science Foundation. For more information on these projects visit our website at www.pblprojects.org.

Introduction

Who this book is for

This is a book of quick and inexpensive investigations for anyone who wants to provide STEM outreach or informal education that focuses on optics and photonics. It is ideal for anyone teaching optics to students. We have used these experiments in both large and small venues with students aged 10 and up as well as senior groups. With the addition of calculations, they were part of a distance-learning community college "home lab" course offered to working technicians and teachers. Because the materials are simple, they can be the perfect answer to a classroom request for a science demonstration or even a science fair project. In some activities, we include sample data tables for recording student observations.

Parents and teachers who are not as familiar with optical physics can use these STEM activities too. Set-up and discussion of most experiments are on the Problem-Based Learning (PBL) Project's YouTube™ channel,[1] which can also be accessed through a link at the bottom of pblprojects.org. Our companion project, *Dumpster*

Optics,[3] has detailed explanations for most of the activities written for parents and teachers, and we provide links to those documents where available. Each lesson also includes PowerPoint™ slides (in English and in Spanish) for presentation to an entire class.

Each educator who uses this book can tailor the activities to the specific audience and age group that they will be working with. For example, what works best as a demo for fifth graders might be better as a small group experiment for highschoolers. Experiment objectives and questions can also be tailored with the specific audience in mind.

Where the activities come from

From 1995 to 2006, the PHOTON Projects of the New England Board of Higher Education were funded by the National Science Foundation's Advanced Technology Education program to develop curricula and materials for teaching optics and photonics to secondary and post-secondary students. The projects also provided professional development to teachers and faculty who, along with industry mentors, developed a community of educators that began in New England and now spans the globe.

The original *Explorations in Optics* (2005)[4] was adapted from some of the favorite demonstrations that PHOTON Projects' participants explored during professional development workshops. Some of the experiments have the designation **OPTICS MAGIC!** This subset of the demonstrations can be used for icebreakers to start a lesson. They formed the basis of most of our workshops for 10 years as we continued to add to and refine our experiments and demonstrations through workshops for students from elementary school through senior citizen groups and at conferences around the world.

By 2015, we realized that dazzling demonstrations of optical magic were great for getting students' attention but not as effective at teaching science. Guided by the U.S. National Science Education Standards (Next Generation Science Standards, NGSS), we developed the longer and more involved *Dumpster Optics* lessons based on these demonstrations and others, but with the requirement that materials had to be commonly found at home or easy to obtain.[5] *Dumpster Optics* was designed to be taught in an upper elementary school classroom as part of a science curriculum. The lessons include PowerPoint slides (in English and Spanish), student worksheets, and detailed parent and teacher notes. If you are looking for more background on optical science than is provided here, *Dumpster Optics* is a good place to start.

During the COVID-19 pandemic of 2020, Optica hosted a series of eight interactive webinars based on the *Dumpster Optics* lessons. Webinar recordings can be found on the PBL Projects' website. The PBL Projects' YouTube channel also hosts, among other educational offerings, videos of most of the Explorations recorded in 2012 with students in the Eastern Connecticut State University's theater program.

A few general hints

You will notice that each activity begins with a question, both to center students' attention and to create curiosity about the activity. We want to encourage active learning and inquiry!

Most of these demonstrations and activities use inexpensive and commonly found materials. All activities have been student and teacher tested, but it's always a good idea to try them out first with the materials you have on hand before doing them with students. Before doing any optics activities, be sure to find out in advance what kind of room lighting you will have. Usually just turning room lights off will work, but sometimes you might need to improvise if an activity works best in darkness. We have found that a deep cardboard box painted black on the inside can double for a dark room in some cases. The goal is to exclude as much room light as possible and to minimize reflecting light inside the box.

Our explanations in the "How it works" sections are aimed at 10- to 12-year-olds, that is, fourth through sixth graders in the U.S. Of course, you can adjust them to the sophistication of your audience, but if you can explain physics to a 10-year-old, you can explain it to anyone.

At the end of these instructions, we provide additional information on materials as well as sources for more "scientific" materials if you prefer to use them. Certain activities require inexpensive but not necessarily household items (for example, chemistry lab watch glasses). These are indicated in the instructions, and sources for finding them are in the Resources at the end of the book.

Notes on Laser Safety

In several of these activities, participants will be working with low-power red laser pointers. Legal laser pointers (<5 mW in the U.S.) are generally safe, but intentional misuse such as staring into the beam can lead to retinal injury. Be careful of lasers purchased on the internet at sites with "awesome" or "wicked" in their names. Some of these are well above the safe level for children's use and, in fact, may be powerful enough to require stringent regulation and oversight by a laser safety officer. Always purchase laser pointers from a reputable vendor.

It's a good idea to discuss rules for laser safety with participants before beginning an activity and, if possible, to write them where all participants can see them. The laser should be kept on a desk or table so that the beam is below eye level. Turn lasers off when not in use, and always be aware of the location of the beam so no one is unintentionally exposed. If you are unfamiliar with laser pointer safety (especially with children), consult the references at the end of this book.[6,7,8]

Chapter 1
The Colors of Light

Exploring Light Spectra

CAUTION! Do not look into the laser cavity or at any reflections of the laser from shiny surfaces. Do not look into the sun!

<u>Question:</u> Can you tell what wavelengths (colors) are produced by a light source by just looking at the light? What kind of instrument do you need to separate the colors to see them?

Materials

- Cardboard tube—a paper towel tube (cut to 15 cm) or toilet tissue tube. For a large group we sometimes buy "craft tubes" that are a bit sturdier (from online sellers such as Amazon).
- Plastic transmission diffraction grating. You can buy one (see References at the end) or make one by peeling the label from an old recordable CD with a piece of tape: scratch the label lightly with the point of a pair of scissors and then pull off the label with tape placed over the scratch. Use sturdy

scissors to cut the CD into a piece that fits the end of the tube. If you buy a diffraction grating, be sure that it's marked "linear."

- Aluminum foil and a rubber band.
- "CAUTION: DO NOT LOOK AT THE SUN OR INTO A LASER" sticker. You can print this on a mailing label or just write it on the tube to remind yourself to be careful.

Science facts

Recognizing colors is one of the first things you learned to do as a child. Isaac Newton used a prism to separate sunlight (or *white* light) into the colors of the rainbow: red, orange, yellow, green, blue, and violet. The colors of visible light are a part of the electromagnetic spectrum,

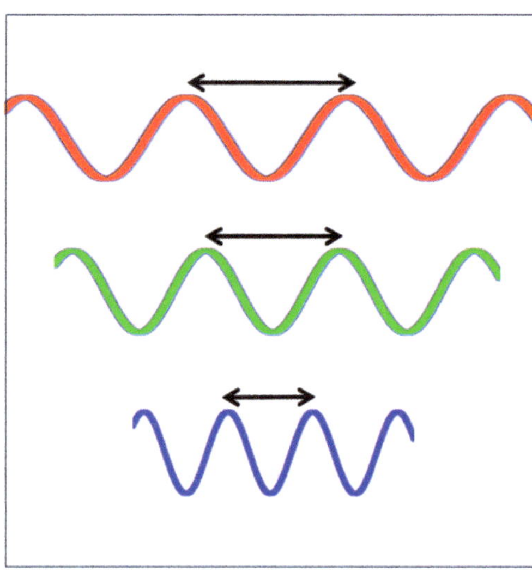

Figure 1.1 Wavelengths (shown by the black arrows) of red (longest), green, and violet (shortest) light.[4]

which also includes radio waves, microwaves, infrared and ultraviolet light, x-rays, and gamma rays.

Each color in the visible spectrum has a different wavelength (the distance from one wave crest to the next). Red light has the longest waves, and violet light has the shortest waves.

Often the color that you see contains several of the fundamental colors of light. For example, a television screen or computer monitor displays many different colors, but if you look at the screen with a magnifier, you will see only very tiny red, green, and blue pixels. The combination of varying amounts of the three colors produces all of the colors that you see on the screen.

In this experiment, you will examine the colors radiated by several different sources of light. When the colors are displayed in order of wavelength, we call it the *spectrum* of the light source. A rainbow is the spectrum of sunlight.

You could use a prism for this experiment, but prisms are a little difficult to work with, and they are often expensive. Like a prism, the very closely spaced tiny grooves of a *diffraction grating* can break light into a spectrum. Diffraction gratings can be very expensive or just a few cents each depending on the quality, but you can also find a grating for free: a discarded recordable CD. Did you ever notice that light reflected from a CD shows rainbow colors? The narrow grooves of a CD are a type of diffraction grating.

To get the clearest spectra, you can construct a *spectroscope*—a device for looking at spectra. (A device that actually lets you measure the spectrum is called a *spectrometer*.) The tube of the spectroscope keeps out room light, and the pinhole or slit on the end lets you look at a narrow "slice" of the source light, so the colors don't overlap.

Procedure

1. Cover one end of a cardboard tube with aluminum foil. Hold the foil in place with a rubber band.
2. Poke a small hole (about 2–3 mm) in the center of the foil with the point of a pencil or, if you can do it neatly, cut a thin slit with a sharp knife.
3. Place the diffraction grating or CD piece on the other end of the tube. We don't usually glue the grating so it can be used without the tube for other experiments, though it's fine to glue the CD when it's in the right position. If you made a slit rather than a hole, the spectrum should be spread out from side to side as in Fig. 1.3. That is, if the slit is vertical when you look into the grating, then the spectrum should be horizontal. If you have trouble

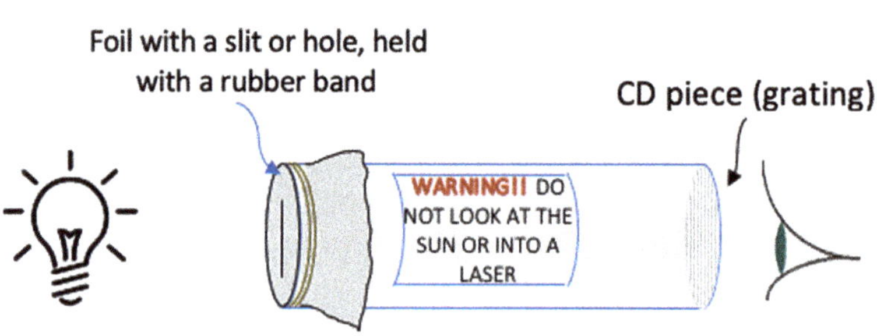

Figure 1.2 The cardboard tube spectroscope. Look through the grating or CD piece on the back end of the tube toward the foil and light source. This drawing shows a slit on the front end, but you can also just poke a small hole into the center of the foil.[4]

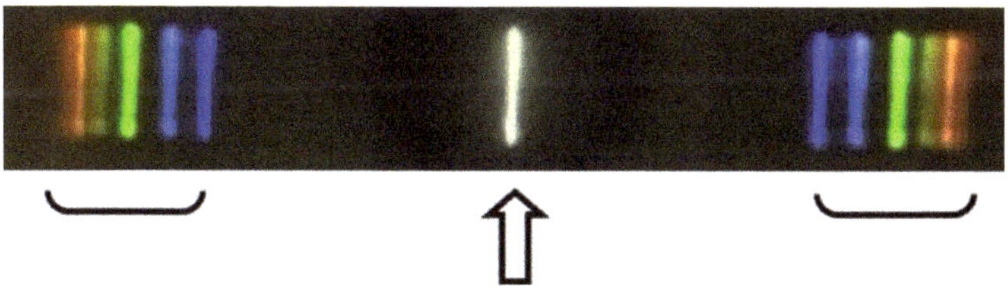

Figure 1.3 Spectrum of a compact fluorescent bulb viewed through a cardboard tube spectroscope. The arrow points to the slit facing the light source. There are two spectra, one on each side of the center. If you don't see them at first, look to the sides of the tube rather than at the end. Sometimes you might see more than one complete spectrum on each side depending on the grating that you use.

finding the spectrum, look on the inside walls of the tube, not at the position of the slit.

Observing light sources

4. Look through the diffraction grating with the pinhole or slit pointed toward the source. You will see a bright spot in the center that will have the same color as the source itself. On either side, you will see one or more spectra—a full rainbow for a source such as a light bulb and individual lines for sources such as streetlights or energy-saving fluorescent bulbs. You may need to look quite far to the side of the central spot to see the spectra. Focus on one of these spectra, which will include some, or all, of the colors of the visible spectrum. For an example of what you will see, look at Fig. 1.3.
5. Find at least three different light sources. Hint: Anything that glows should work, for example, light bulbs and LEDs, a cell phone screen, a

candle flame, streetlights, etc. A colored light, such as a green night-light or a colored holiday bulb, would be a good choice.

6. If you have a laser pointer, you can include it as one of your sources. However, DO NOT look directly into the laser with your spectroscope. Shine the laser on a piece of white paper and look at the light reflected from the paper. You can view the spectrum of sunlight the same way by looking at sunlight reflected from white paper.

Recording your observations

Fill in the data table with one line for each light source. List the types of light that you were observing (LED, laptop screen, incandescent bulb, candle, etc.) and write a description of the spectrum you see through the spectroscope.

Observations

Type of light source	Colors you saw in the spectrum and which color was brightest

Conclusions

- Do the spectra of the colored sources contain only the color you see with your eye? That is, if you look at a red light bulb, will it always show only red light? Or might other colors be present?
- Do some light spectra have more colors than others?
- Why do you think the spectra of different light sources are different?
- Can you tell what colors are in a light source by looking at it with your unaided eye?

Application

Spectroscopy has many uses in astronomy, earth science, food inspection, crop health, manufacturing, sensing, and more. For example, Mars rocks can be studied using a *laser induced breakdown spectroscopy* (LIBS) instrument. A Mars rover equipped with LIBS fires a laser at a rock, making a small puff of glowing vapor that is imaged by the spectrometer. The pattern of wavelengths produced tells scientists what the rock is made of. You can find more interesting applications for studying both Earth and space on the NASA website.[9]

OPTICS MAGIC!

What Color Is a Tomato?

<u>Question:</u> Can your eyes be fooled by the color of a light source? Can you make your friends think a tomato is a plum?

Materials

- A small tomato that you can mostly hide in your hand, for example, a cherry or grape tomato (or a few pieces of colorful candy such as Skittles® or M&M's®).
- At least two different colored lights, such as finger LED lights or holiday lights. You could also use a flashlight, covering the end with blue, green, or red plastic film or a tightly stretched balloon, but the light will be quite dim, requiring a dark room.

Science facts

The color you see when you look at an object depends on the wavelengths reflected by the object, the wavelengths present in the light illuminating the object, and the color sensitivity of your eyes. A red tomato reflects a range of wavelengths, primarily red but also extending into orange.[10] However, the skin is shiny so when it's illuminated by a blue LED that blue light is reflected (the LED light doesn't contain any red wavelengths). If there is some stray white light in the room coming from the windows or other lights, the red content of that light is also reflected by the tomato—and the tomato looks like a purple plum due to the combination of both red and blue light.

Procedure

1. In a dark room, hold the tomato in your hand so that only a small portion of the surface is visible. (You don't want the shape of the tomato to be visible, just a bit of the skin.) If you are unable to darken the room, place the tomato in a deep box so that it is well shaded from ambient light. It helps to paint the inside of the box flat black to minimize reflections of room light.
2. Shine the colored light on the object and observe the color of the illuminated surface. For example, a red tomato under blue light looks like a purple plum! Try different colors if you have them, and different objects as well.

Record your observations

For each object and each illumination type, answer the following questions. Be sure to say what the object was and what type of illumination you used.

1. What color is the object under white light (sunlight)?
2. What color light did you use?
3. What color did the object appear under this color of light?
4. Why did the object appear to be the color you saw?

Observations

Object you observed	Color of light used	Color the object appeared	Why did the object appear to be the color you saw?

Application

Lighting plays an important role in marketing. Figure 1.4 shows the effect of illumination on the same objects.

Figure 1.4 Photographs of the same objects taken with (*left*) warm (2700 K) and (*right*) daylight (6500 K) illumination.

The items are identical in both the left and right photographs, so why do they look different? Check out the lighting in a local supermarket. Are the same lights used for meat and produce?

Chapter 2
Exploring Pinhole Images

<u>Question:</u> **Can you make an image (picture) with just a cardboard box and a pinhole?**

Materials

- Large carton or box
- Aluminum foil
- Sharp pencil (or a thick needle, such as an embroidery needle)
- Tape
- Translucent paper, such as waxed paper or tracing paper
- Electrical tape to cover any light leaks, if needed

Science facts

How does a pinhole camera work? As you can see in Fig. 2.1, sunlight reflected from the top of the tree passes through the small pinhole in the front of the box and strikes the back of the box at the bottom. Light from the

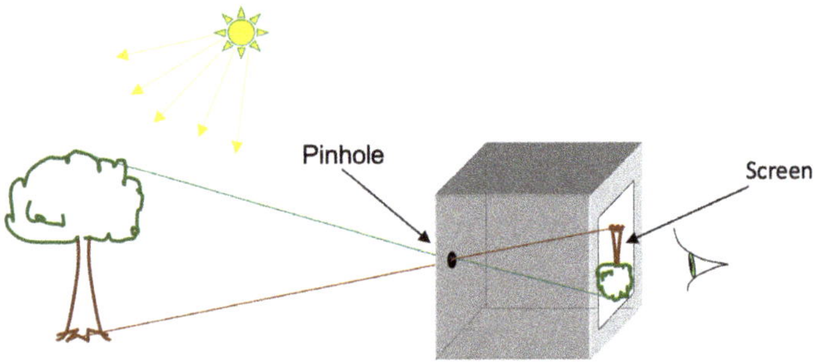

Figure 2.1 The pinhole camera viewer uses a tiny pinhole. Geometry shows how light from different parts of the object (tree) are kept separate to form an image.[4]

bottom of the tree also passes through the pinhole and strikes the back of the box at the top. The light from the top and bottom does not overlap because the pinhole is so small. Thus, an image of the tree is formed on the back of the box. The image is upside down, and its size can be found by geometry (similar triangles).

If the hole is too large, the overlapping rays will form a blurry image or no image at all. If it's too small, the image will be very dim because so little light enters the box. To use the box as a camera, it must not have any light leaks that allow in more light than the pinhole does. The film is placed on the inside, at the image location. Exposure times can be very long—several seconds outside on a sunny day to up to several hours indoors. The Resources chapter has links to instructions to make and use a simple pinhole camera.[11]

Procedure

1. Cut an approximately 5-cm square hole in the center of one end of the box. Check to make sure that the edges and corners of the box will not let in any stray light. You can cover any small holes or cracks with black electrical tape if necessary.

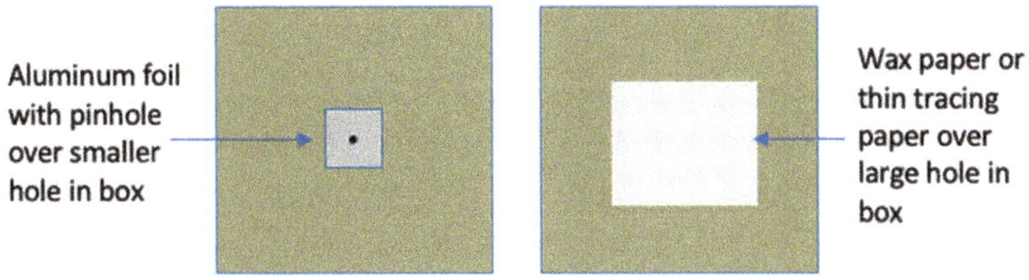

Figure 2.2 Front (*left*) and back (*right*) of the box.

2. Cut a larger hole, around 10–15 cm square, in the box opposite the pinhole. The actual size will depend on the size of the box.
3. To make the pinhole, stack a few 6- to 7-cm square pieces of aluminum foil cut slightly larger than the smaller hole in the box. Pierce the stack with the tip of a sharp pencil or needle. The inner foil pieces should have neat pinholes with clean edges. The pinhole should be a couple of millimeters across depending on the size of the box. Tape one of these foil pieces over the hole in the box, centering the pinhole (Fig. 2.2).
4. Cut a piece of waxed or tracing paper to cover the large hole. This will be the viewing screen.

Aim the pinhole toward a light source and look at the screen at the back of the box. A lamp in a darkened room makes a good object. Or stand in a darkened room and point the pinhole toward a sunny window.

Alternative method of viewing the image: This requires a fairly large box (you need to put your head almost inside it!), but it gives a sharper image. Instead of a waxed paper screen, use the inside of the box to see the image. To do this, do not cut a hole in the back of the box. Leave the bottom of the box open and hold the carton over your head. Look for the image on the inside of the box facing the pinhole. This works best with a bright light bulb as the object.

Observations

What did you use for an object? Describe the image you saw. Was it upright or upside down? Sharp or blurry? Describe any other features of the image that you noticed.

Application

Many photographers enjoy creating pinhole images because of their unusual qualities. Cameras with laser-drilled pinholes can be bought online, or cardboard cartons and oatmeal containers can be used to make simple "recycled junk cameras" that can be used with film paper for quickly developed black and white photos.[11] Figure 2.4 shows photographs taken with an oatmeal box camera. The film is curved and produces oddly curved photos.

Figure 2.3 An oatmeal box pinhole camera. The pinhole is in the piece of aluminum taped to the box. The loose piece of black tape is the "shutter."

Figure 2.4 *Top*: The film paper was taped inside the box, following the curve. The image on film paper is a negative. *Bottom*: The image was scanned and digitally processed to form a positive image. The curvature of the film in the box makes it look like the wires are curved. (Bottom photograph courtesy of Photonics Media.)

Figure 2.5 *Top*: Pinhole images of the sun under dense vines.[3] *Bottom*: Pinhole images of the sun under a leafy tree during a partial eclipse of the sun.

A very common (but often unrecognized) pinhole image can be seen under the canopies of leafy trees. The round blotches on the ground are pinhole images of the sun; these round spots turn to crescents during a solar eclipse (Fig. 2.5).

https://t.ly/lVXYP

You can find a video of how to make a pinhole viewer with everyday household items at the QR code and link above.

Chapter 3

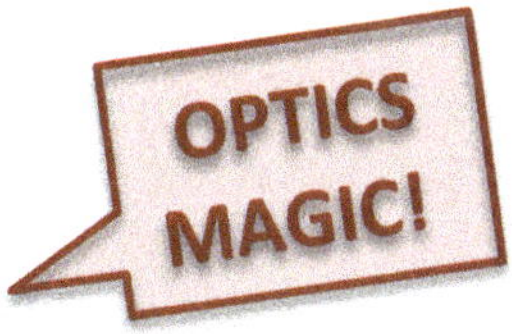

Exploring Reflection

The Disappearing Beaker

Note: This is traditionally a messy experiment but amazing and fun. A less messy variation is also described in the procedure.

Questions: What does transparent mean? Can you see a transparent object? Can you make a glass disappear?

Materials

- Two Pyrex® (borosilicate glass) laboratory beakers or custard cups, a small one that fits completely inside the larger one. Other types of glass may or may not work—experiment to find out! The inner glass is the one that is important—a small Pyrex® cup inside a different glass container would work.
- Inexpensive vegetable oil (really—don't use that expensive olive oil here)

Science facts

In order for you to see an object, light must go from the object into your eyes. A glowing object such as a light bulb emits light that you can see directly, but other objects must reflect light from the sun or another light source for you to see them.

Whenever light travels from one medium to another (for example, from air to glass or glass to water), part of the light is reflected, and part of the light is transmitted into the second material. The amount of reflected light depends on the *speed of light* in the two materials. Scientists speak of the *index of refraction* of a material, which is a way to express how fast light travels. For example, glass has an index of refraction of around 1.5, which means that light travels 1.5 times faster in vacuum than in glass.

When light goes from air to glass, it slows down in the glass, and about 4% of the light is reflected at the surface. That's enough for you to be able to see the glass, even though we say it's transparent. This is illustrated in Fig. 3.1. But cooking oil and borosilicate glass have about the same index of refraction—that is, light travels at the same speed in both. So, no light is reflected if the glass is surrounded by oil, and therefore you can't see it!

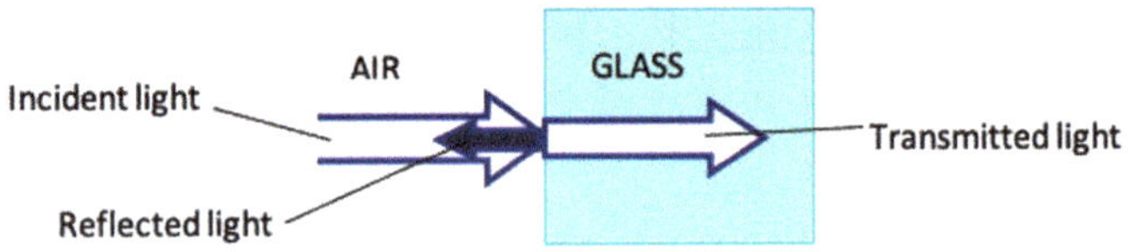

Figure 3.1 Light is reflected from glass because it has a different index of refraction than air. There will be a reflection at the second glass surface too, but this is not shown in the drawing.[4]

Procedure and observations

1. Place the smaller beaker in the larger beaker.
   - Can you see the smaller beaker?
   - Why or why not? (Think about the speed of light in air and in glass.)
2. With the smaller beaker still inside the larger beaker, carefully pour the oil into the smaller beaker only until it is full (see Fig. 3.2).
   - Can you see the smaller beaker?
   - Why or why not? (Again, think about the speed of light.)
3. Now continue pouring the oil into the smaller beaker so that it overflows into the larger beaker. Continue pouring until the smaller beaker is completely submerged in oil.

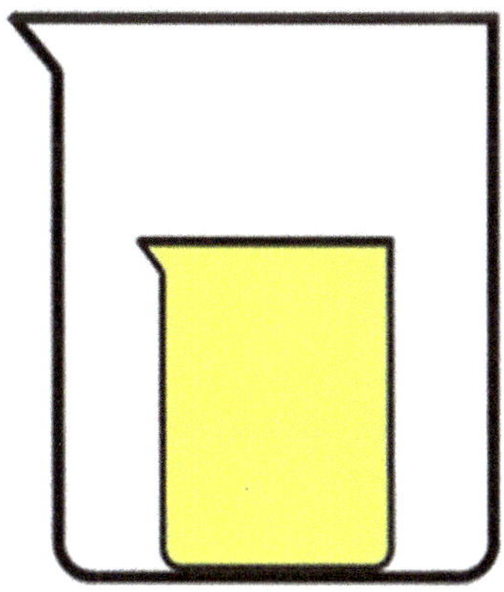

Figure 3.2 Smaller beaker with oil inside a larger beaker.[4]

- Can you see the smaller beaker?
- Why or why not? What does this mean about the speed of light in the oil compared to its speed in the glass?

Alternative procedure

This demonstration may also be done with water (and less mess) using gel beads from a craft store. Look in the flower-arranging aisle for colorless gel beads, also called hydrogel spheres, to use in water. When saturated with water, the beads can't be seen. It does take a while for them to fully expand, however, so put them in water several hours (or the night) before a demonstration. The beads are invisible because light doesn't change speed going from water into the beads (which are nearly all water themselves), so there is no reflection. But if you put your hand in the glass of watery beads, you can feel them!

If you want to dry the beads to reuse them, you need to expand them in distilled water, or they will turn brown as they dry.

Application

To minimize the amount of light reflected (and maximize the amount transmitted) at a surface between two materials, the index of refraction (speed of light) in both materials should be as close as possible. Sometimes, *index-matching fluids* are used, for example, when two optical fibers are joined in a temporary splice to minimize reflection back into the signal source.

An application that you may have seen is the gel applied before a medical ultrasound examination. Many people think the purpose is to make the ultrasound transducer slide more easily over the skin. In fact, it is index-matching gel to maximize sound energy transmission from the transducer into the body. Without the gel, much of the ultrasound would be reflected by the thin layer of air between the transducer and the skin. To get a good echo return signal from inside the body, it's important that most of the sound be transmitted, not reflected, at the skin's surface.

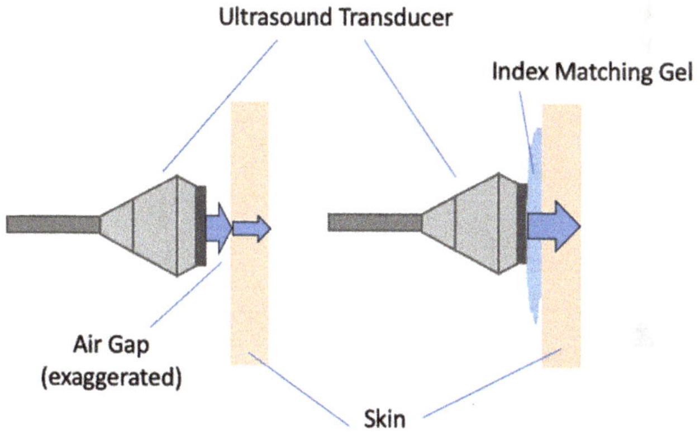

Figure 3.3 *Left*: Sound waves traveling from left to right are partially reflected at each surface where the speed of sound changes as sound goes from the ultrasound transducer through a thin layer of air and then into skin. (Reflected sound is not shown.) *Right*: Index- (of refraction) matching gel between the ultrasound transducer and skin minimizes reflection and maximizes the amount of energy entering the body for a better image.

Laser Bullseye and the Law of Reflection

CAUTION! Do not look into the laser or at any reflections of the laser from shiny surfaces.

Question: How does reflection work? Can you use the law of reflection to hit a target with a laser beam?

Materials

- A laser pointer, placed on blocks or books as needed to get the beam to mirror height
- A mirror. You can use CD jewel cases for mirrors. Remove the inside label and replace it with black construction paper. They even stand upright by themselves! You can also use a small glass or plastic mirror, but you need to find a way to make it stand vertically. With "real" mirrors, you can have the laser strike more than one mirror. (A CD case doesn't reflect well enough for that.)
- Protractor (printable ones from the internet are fine—these help to measure angles)
- String (this helps with layout)
- Meter stick (this makes sure everything is far enough apart)
- Target (see Fig. 3.6)

Science facts

This activity is based on an actual test given to laser technicians as part of a job interview! In that case, the laser had to strike many (very expensive) mirrors and hit a very small target. In this case, we'll use one CD reflector (or one or more real mirrors) and try to hit the target.

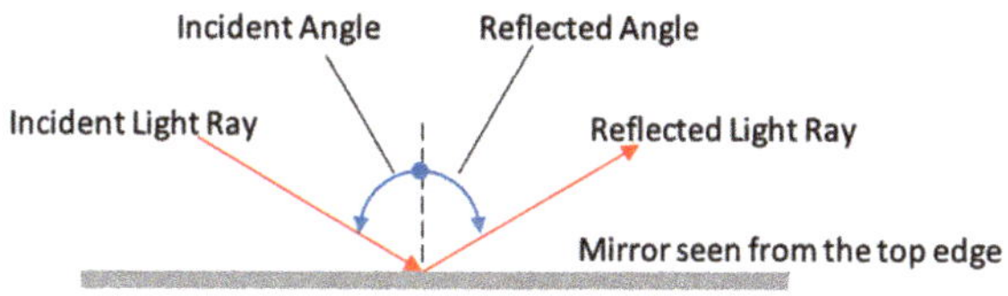

Figure 3.4 The law of reflection. We are looking down from the top at a mirror that is standing vertically on a table. Notice how the angles are measured—from a line drawn perpendicular to the mirror, a so-called *normal* line, not from the mirrored surface itself. Normal means perpendicular. The light rays (red arrows) show how a laser beam would be reflected from the mirror.[4]

When light strikes a shiny surface, it leaves the surface at the same angle it entered. We call this the law of reflection:

"The angle of reflection equals the angle of incidence."

The light that strikes the surface is *incident* and the light that leaves is *reflected* (see Fig. 3.4). In this Exploration, you will use this law to help you place a reflective CD case (or one or more mirrors) so that a laser beam hits a target.

Rules

1. The laser and target will be set up for you by the instructor or assistant. *You may not move the laser or the target.*
2. The instructor will decide how much time to allow and if there will be multiple rounds.
3. **The laser must be kept turned off until you have set up the mirrors.** When you are ready, the instructor or assistant at your table will turn on the laser. *You may not move any mirrors while the laser is on.*

4. If you don't get 20 points on your first try, the instructor or assistant will turn the laser off, and you may move the mirror and try again. You can keep trying until time is up for that round.
5. If the laser falls on a line between regions of the target, you will be awarded the average of the points on either side.
6. The mirror must be at least 25 cm from the laser. (You can make this distance larger if you have space or want an additional challenge.) If you use multiple mirrors, they need to be 25 cm apart also. See Fig. 3.5 for an example setup of the laser bullseye.

Application

There are many applications requiring a laser beam to be "steered" or directed in one direction or another, often changing direction at very high speed. Often this is done with motorized fast-steering-mirror systems. One example is a laser light show, where small-area beams trace out designs that appear to stand still or move slowly in the air. Can you think of other examples?

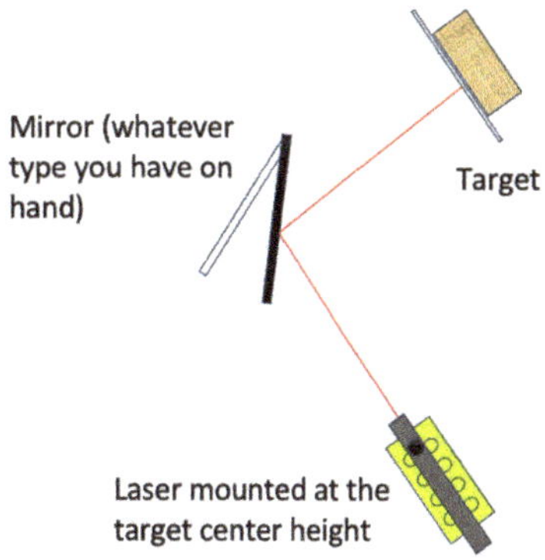

Figure 3.5 *Left*: Diagram of the laser bullseye. *Right*: Setup using a laser pointer on plastic brick stack, a CD jewel case mirror, and a target, also mounted on a CD jewel case. Of course, you can use whatever mirrors and mounts you have handy! You can also make the target larger or smaller.

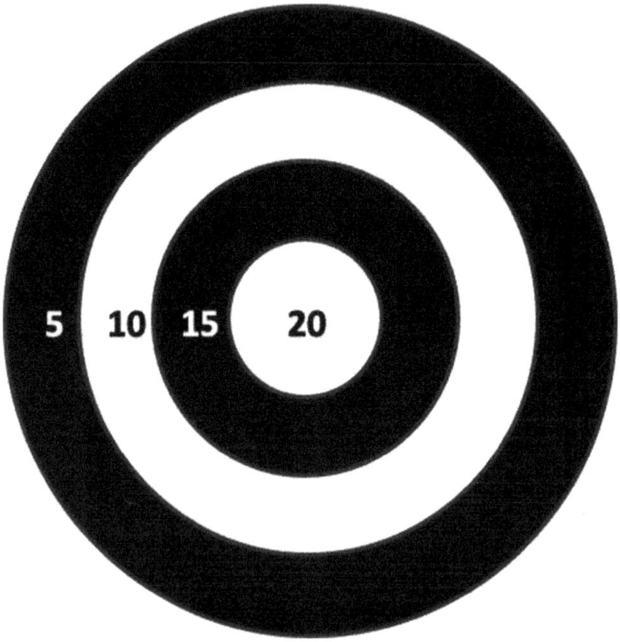

Figure 3.6 Sample target. Scan and cut out the target, glue or tape it to a piece of cardboard, and mount it on a block so that it is vertical. You can also create a target of your own.[4]

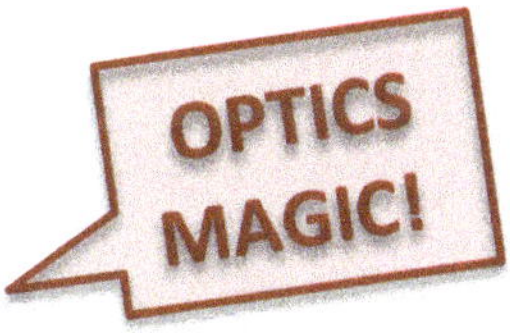

Giant Bedazzled Kaleidoscope

<u>Question</u>: How many times can you multiply yourself?

This is a wonderful activity that is actually less expensive than making individual kaleidoscopes for a whole classroom. Plastic mirrors are usually quite expensive, and glass mirrors are fragile, so this is a good compromise.

Materials

- Three large rectangular mirrors (we use the kind you mount on the back of a door or a wall, available inexpensively at big box stores)
- Duct tape, lots of it!
- Optional: Decorations for the backs of the mirrors

Science facts—reflection from more than one mirror

Kaleidoscopes work by reflection, and the law of reflection is not difficult to understand. But the explanation for the formation of multiple images is not as simple. Multiple reflections can be easily illustrated, though, with two small, flat square or rectangular mirrors. Glass mirrors can be hard to find, and they might be too fragile for some students. CD jewel cases can be used as mirrors that work nearly as well. Remove the liner notes (if any) and replace them with black construction paper. It's easy to stand the cases vertically without needing to resort to blocks or other supports.

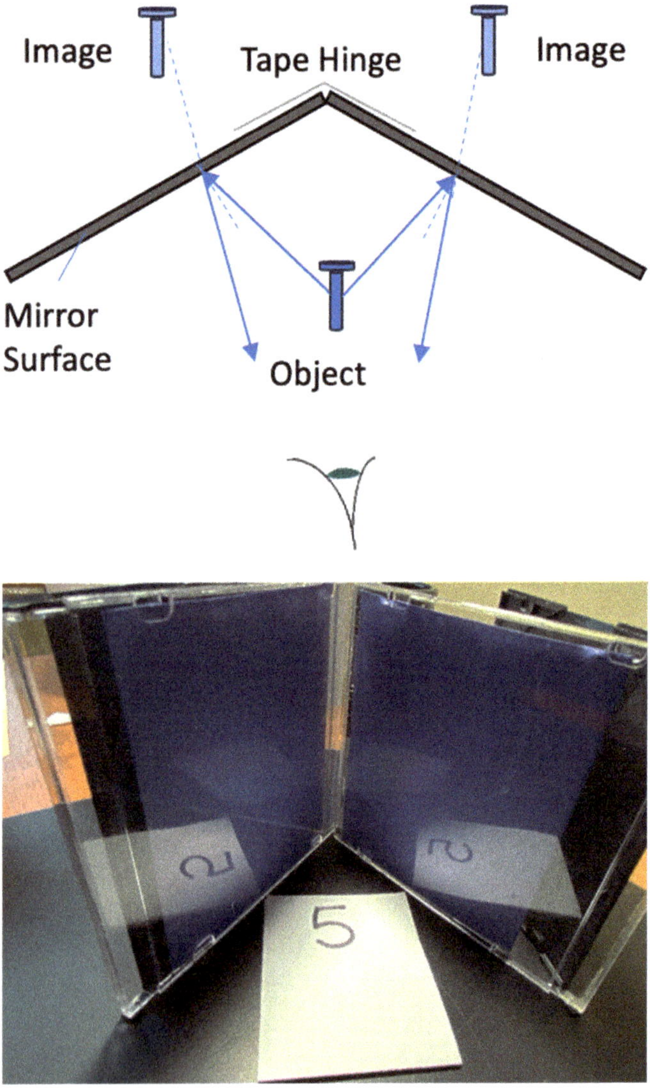

Figure 3.7 *Top*: Diagram showing two images formed by hinged mirrors at about 120 deg. *Bottom*: Two CD jewel case mirrors showing four reflections of the number five placed between them.

Place the two mirrors so they stand vertically making an angle of around 120 deg. You can prop up separate mirrors with small wood blocks or Lego® bricks. (It helps to have a "hinge" of tape to hold the mirrors together.) Place a small object such as a paper clip between the mirrors. As shown in Fig. 3.7, an image forms in each of the mirrors, as expected. As the angle between the mirrors is made smaller, more images appear due to multiple reflections.

An interesting exercise is to count the number of images that appear as the angle between the mirrors is made smaller. Use a printed number "5" for the object and observe the orientation of the multiple reflections—do they all face the same way? How many images can you make as the angle between mirrors gets smaller? Can you find a relationship between the number of images and the angle between the mirrors? Note that the additional reflections are difficult to see with a CD jewel case mirror; they are much brighter with "real" mirrors.

Procedure and observations

1. Carefully tape the three wall mirrors together along their long edges as shown in Fig. 3.8. You can also tape around the edges of the end openings to protect them from breakage if they are exposed. The shiny mirror surfaces should be facing inward.
2. Optional: Decorate the outside (backs) of the mirrors with jewels, stick-on sparkles, etc. This really grabs attention.

3. To use: Have students stand on each end and look into the kaleidoscope opening.
   - How many faces do you see?
   - Do you see complete faces or parts of faces?
4. Try looking at other objects such as a large number "5" (to see the orientation of reflections) or brightly colored pictures.

Application

You often see an application of multiple images when you are in a fitting room trying on clothes. But kaleidoscopes are interesting in their own right. Since its invention by Sir David Brewster in 1816, the kaleidoscope has been enthusiastically embraced by both curious children and serious artists. The Brewster Kaleidoscope Society was founded in 1986 "to share and promote the beauty, creativity, and joy of these mirrored tubes of magic." The society's website has links to kaleidoscope artists in the U.S., U.K., and Japan as well as diagrams showing how the number and orientation of images change as the angle between the mirrors change.[12]

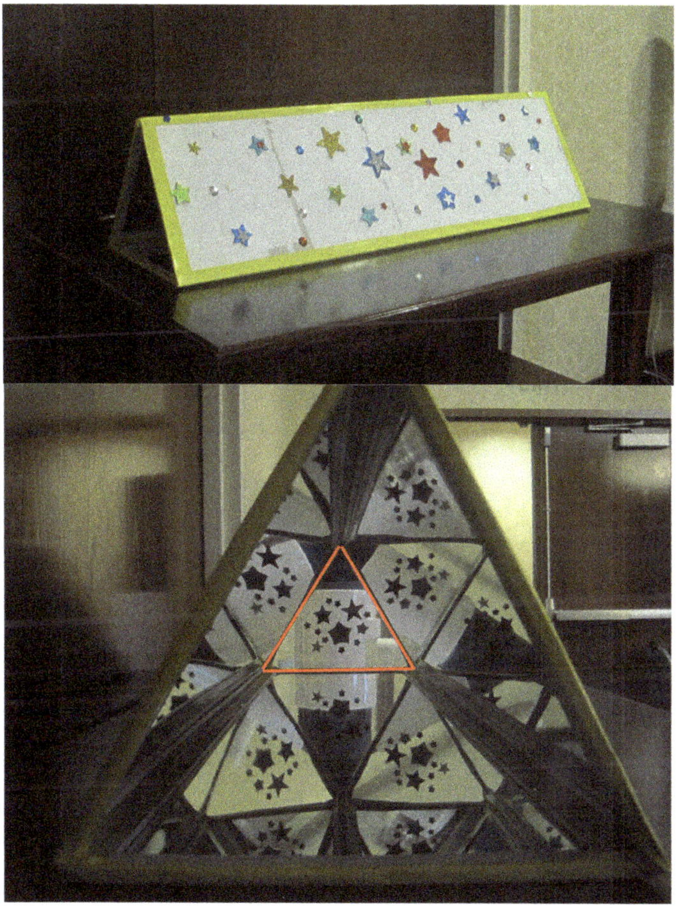

Figure 3.8 A Giant Bedazzled Kaleidoscope. *Top*: Three large door mirrors (4 ft or about 120 cm on the long edge) are held together by yellow duct tape. The tape is also applied around the two open ends. *Bottom*: Looking inside the kaleidoscope from one end. A clear plastic sheet decorated with star stickers is held at the far end. In the photograph, the far (open) end of the kaleidoscope is outlined in red.

Chapter 4
Exploring Refraction

Exploring Refraction with Gelatin

CAUTION! Do not look into the laser or at any reflections of the laser from shiny surfaces.

Questions: How does refraction work? Can you focus light with gelatin?

Materials

- Slabs (around 1.5 cm thick) of very stiff gelatin. You can use plain gelatin or flavored, sugar-free gelatin. It's best not to use the kind with sugar because it makes a sticky mess. See preparation instructions in the procedure.
- Something to cut with—knives work, but 2-cm-wide strips cut from a plastic folder can be bent to shape and are smooth and less dangerous. Large round cookie cutters are also good for making "lenses."
- Laser pointer.
- Ruler and protractor if you want to take measurements.

Science facts

In vacuum, light travels at 300,000,000 meters per second or 186,000 miles per second. When light enters a transparent medium such as glass (or gelatin), it slows down and its wavelength (the distance between wave crests) shortens. Because of this, a beam of light traveling from one substance to another can change direction.

To understand why a wave changes direction when it changes speed, think about a marching band in neat even rows, marching along a paved surface bordered by deep mud. The marchers can walk faster on the pavement than in the mud, where their feet sink in the ooze. When the marchers at the end of each row reach the mud, they will take shorter steps and slow down. The result is that the rows will bend. (See Fig. 4.1.) In the drawing, the white circles represent the heads of the marchers.

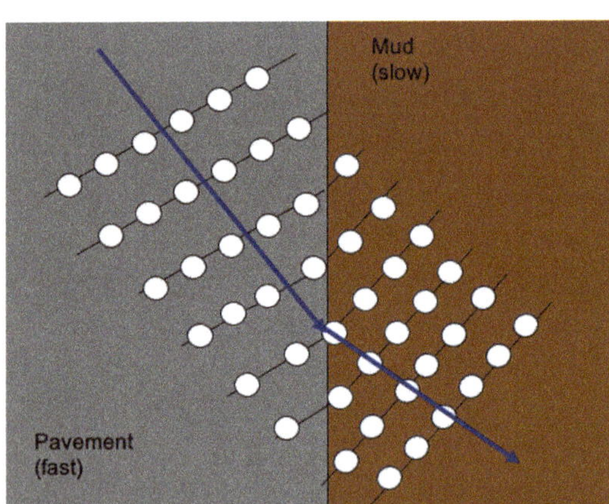

Figure 4.1 The rows bend when the marchers go from pavement (fast steps) to mud (slower steps).[4]

Waves behave in a similar fashion. In Fig. 4.1, the rows of marchers are like the crests of waves. You can see how waves will bend when they go from a medium where they travel fast to a medium where they move more slowly, for example, if waves go from air into water or glass.

The bending of waves as they go from one medium to another is called *refraction*. Refraction of light explains many things you see around you every day, such as the apparent bending of a spoon when it is partially submerged in water and how lenses work, including those in eyeglasses. When scientists talk about refraction, they speak of *incident* light, the light that strikes the surface, and *refracted* light, the light that enters the second *medium* (or material). The angles of incidence and refraction are measured from a line drawn at 90 deg to the surface. This is similar to the law of reflection where there were incident and reflected rays, both measured from a line drawn perpendicular to the surface (Fig. 4.2).

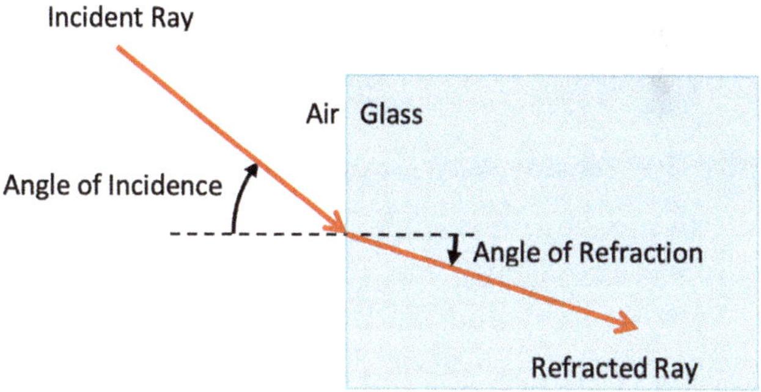

Figure 4.2 Incident and refracted rays, and angles of incidence and refraction. Note the way the light bends when it goes from air on the left to glass on the right. Compare this to the bending of the rows of marchers in Fig. 4.1. What will happen when the light gets to the second surface on the right? Which way will the light bend?[4]

Procedure and observations

1. **To make the gelatin:** Make the gelatin in a pan that will allow the gelatin block to be around 1.5 cm thick. If it's too thin, it will be hard to handle. Use half of the usual amount of water recommended on the package. On flavored gelatin packages, follow the recipe for "blocks" or "Blox" as they are called on the Knox® package. These are also referred to as Jigglers® on the Jello® package. Remember to lightly oil the pan to make the gelatin easier to remove. If it is stiff enough, it doesn't need refrigeration except on really hot days.
2. To make a large quantity of gelatin for experimenting, try asking the local butcher for small plastic meat trays. It's easier to handle in small trays rather than in a large pan.
3. Carefully remove the hardened gelatin and cut it into shapes for experimenting. Use a cookie cutter or a thin flexible piece of plastic to cut shapes as described below. A knife is not recommended unless it has a thin, smooth (not serrated) blade.

Activity 1—Which way does light bend?

4. Cut a gelatin rectangle about $3\ \text{cm} \times 4\ \text{cm}$. The size is not important, but two opposite edges must be very straight and smooth. You can cut it with a non-serrated knife or slice it with a plastic strip or a thin, smooth ruler. On a piece of paper, draw two long lines that cross at a 90-deg angle. Draw a few more lines for different angles of incidence (see Fig. 4.3) and place one of the straight edges along a line as shown in the figure.

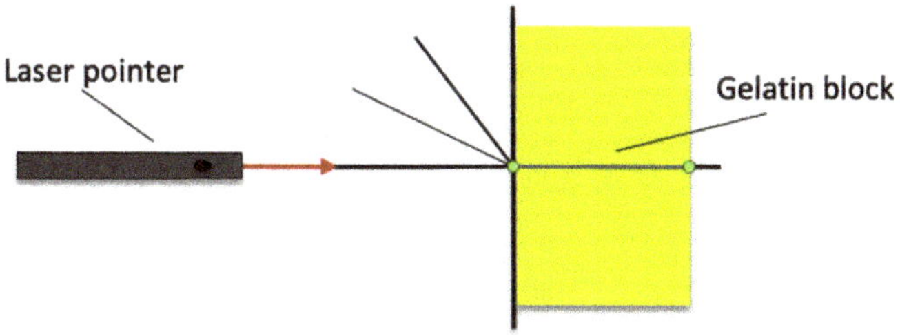

Figure 4.3 Gelatin block and paper with lines drawn on the paper below, as seen from above.[4]

5. Place the laser so that it strikes the gelatin block straight on (at a 0-deg incident angle, as shown in Fig. 4.3).

   Does the beam of light bend when it goes from the air straight into the gelatin? Or when it goes from the gelatin back into the air on the other side?

6. Use the laser pointer to shine the beam along each of the other lines you have drawn. Each time, notice which way the beam bends as it enters the block of gelatin.
7. **Observation:** Which way does the beam bend as it enters the gelatin block? As the angle of incidence increases, does the beam bend more, less, or the same amount?

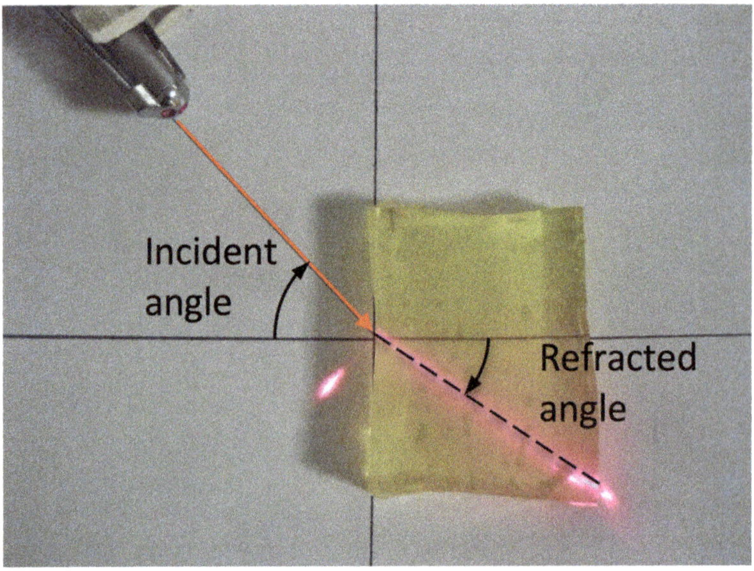

Figure 4.4 Photograph of the gelatin block and laser. The laser beam is highlighted so that it is easier to see. In fact, light from a laser pointer isn't normally visible at all in air—do you know why?[4]

Activity 2—Make a lens!

8. To cut gelatin "lenses" into the shapes shown in Fig. 4.5 you will need to use the circle cookie cutter and a flat block of gelatin. Notice that the curved surfaces of the lenses are actually parts of circles. Practice until you can make shapes with smooth sides that look like the ones in Fig. 4.5. The lens on the left has two convex (bulging outward) sides and the lens on the right has two concave (curving inward) sides.

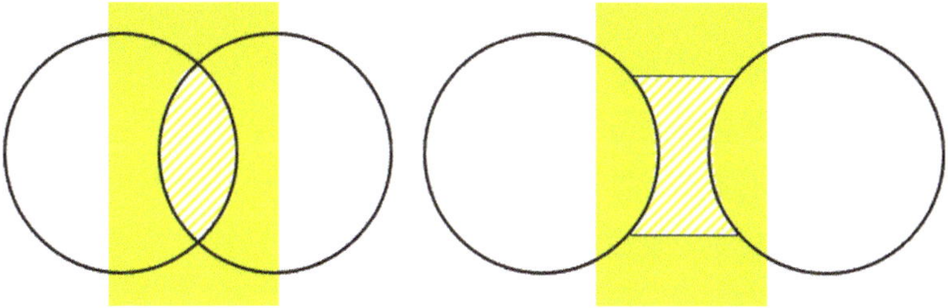

Figure 4.5 Lens shapes cut out of a gelatin block using a circular cookie cutter. The circles show the positions of the cookie cutter. The light-yellow shaded shapes are the gelatin "lenses" that you will use for this experiment.[4]

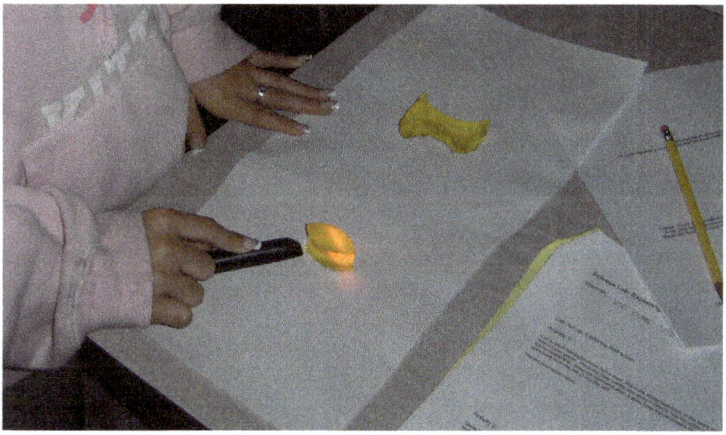

Figure 4.6 Fifth-grade student experimenting with converging and diverging lenses cut from gelatin.

9. Place one of the "lenses" on a piece of paper. Shine the laser through one edge of the lens and observe how the light behaves as it passes through the lens. Shine the laser pointer along one of the arrows, as shown in Fig. 4.7, and trace the path of the ray on the other side with a pencil.

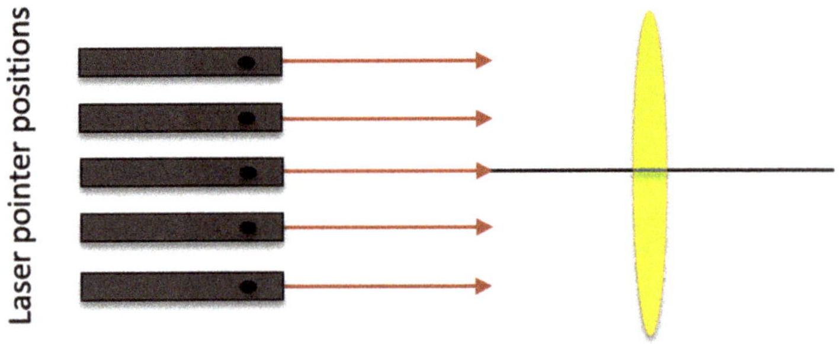

Figure 4.7 Several positions of the laser pointer as described in Step 10. The rays enter the lens parallel to the line through the center of the lens. Trace the path of the ray with a pencil to show its direction after it passes through the lens (not shown).

10. Move to another (parallel) position and repeat. Draw at least five rays and see what happens on the other side of the lens. A lab device called a *ray box* makes this super easy since all of the rays can be seen at once (with no drawing required), but they are more expensive than a dollar store laser pointer. (See the supplier list in the Resources chapter.)
11. Repeat with the other shaped lens.

What does the lens on the left in Fig. 4.5 (called a converging lens) do to the light? Notice that this lens has convex surfaces on both sides so it is thickest in the center.

What does the lens on the right in Fig. 4.5 (called a diverging lens) do to the light? This lens has concave surfaces on both sides so it is thinnest in the center.

CHALLENGE: Can you make a converging lens that is convex on one side and concave on the other?

Remember, it just needs to be thicker in the center than on the edges. This is called a convex (or positive) meniscus lens.

Conclusion

How does the shape of the lens affect the way light travels through it? Use what you know about refraction to show how one lens causes light to focus, and one causes light to spread out.

If the lenses were made of another material, say glass, that bends light more than gelatin, how would the lenses behave differently? Use a diagram to explain.

Applications

The most obvious application of refraction is lenses, found in everything from eyeglasses to cameras to telescopes. But refraction effects are much more common than just in lenses; for example, judging the depth of a fish under water or explaining atmospheric phenomena such as mirages requires the principle of refraction. Challenge students to see how many applications of refraction they can think of! You can find more refraction experiments in the Dumpster Optics lesson, "Bending Light—Refraction."[3]

Gelatin Optical Fiber

OPTICS MAGIC!

Questions: How does optical fiber work? Can you make an "optical fiber" out of gelatin?

Materials

- Slabs (around 1.5 cm thick) of very stiff gelatin. See the Gelatin Optics exploration for details and instructions.
- Something to cut the gelatin—you need to make a long, straight, smooth cut. A strip cut from a plastic folder would work.
- Laser pointer.
- Advanced topic: Sugar if you want to try GRIN (graded index) gelatin.

Science facts

If you tried the gelatin refraction exploration, you saw how light bends as it goes from air to gelatin and then back into air. When light goes from where it travels faster to where it travels slower, it bends toward the line perpendicular to the surface (the normal line). If light goes from where it travels slower to where it travels faster, it bends away from this line. Light travels faster through air than it does through gelatin.

But under certain circumstances, light in the gelatin is "trapped" and can't exit into the air. This is called *total*

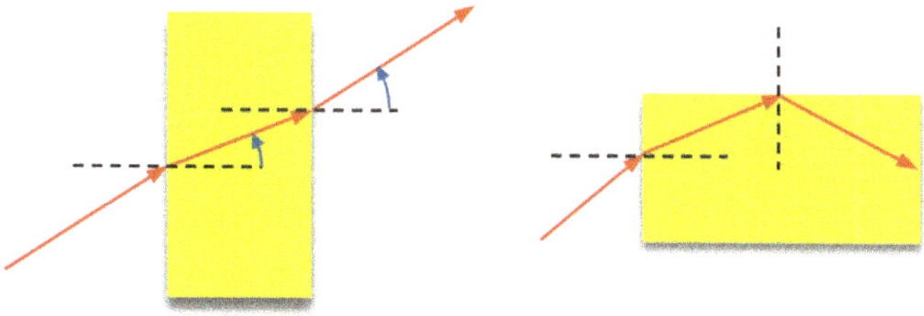

Figure 4.8 *Left*: Refraction of a light ray as it goes from air to gelatin (yellow) and back into air. Light bends toward the normal line when it enters the gelatin from air; it bends away from the normal line when it exits the gelatin into air. *Right*: Inside the gelatin, light strikes the air–gelatin surface at a large enough angle to be reflected back into the gelatin rather than travel into the air.

internal reflection, and it happens when light traveling in the gelatin meets the air surface at a large enough angle (called the *critical angle*). When this happens, all of the light is reflected back into the gelatin instead of being transmitted into the air. You may have seen this effect looking into the corner of a fish tank—it looks like there are duplicate fish because the glass and water in the tank act like a mirror.

Optical fiber works by trapping light by total internal reflection. Ultra-pure glass allows light to travel for many tens of kilometers before it requires amplification. It's difficult to believe, but hair-thin optical fiber can transmit much more data than copper wire, which is why it rapidly replaced copper as a transmission medium for communications. Optical fiber has many other uses as well, such as for sensors and image transmission (like for endoscopes).

Procedure and observations

1. Cut the longest thin strip you can out of your slab of gelatin. It should be at least 1 cm wide so that you can handle it without breaking.
2. Shine the laser at an angle into one end.
   - What does the light do inside the gelatin?
   - If you change the angle (move the laser), what do you notice?
   - Do you see any light exiting the far end?
3. Try bending the "fiber" into a gentle curve.
   - Does the light follow the "fiber" around the bend?

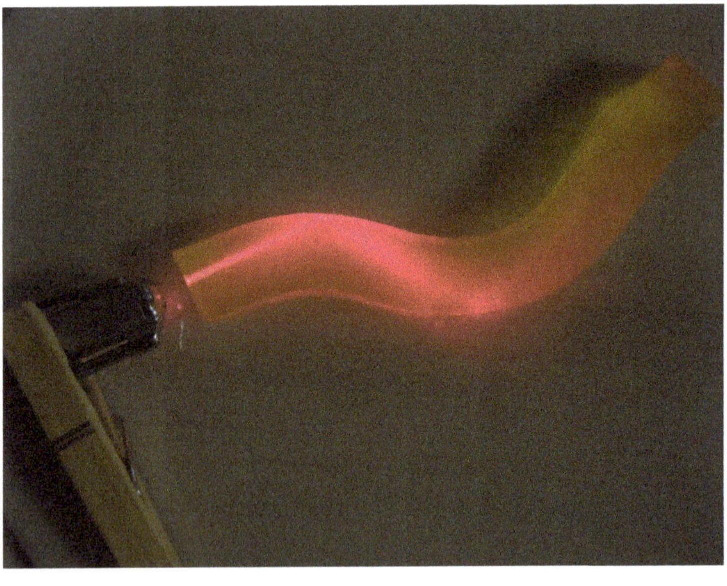

Figure 4.9 A gelatin "optical fiber" showing total internal reflection. The laser pointer is held to the "on" setting by a clothespin.[3]

Advanced topic: GRaded INdex gelatin

Graded-index materials have an index of refraction that changes gradually throughout the material rather than abruptly as with the gelatin–air boundary. The index of refraction of gelatin can be increased (the speed of light decreased) by adding sugar.

Procedure

1. To make GRIN gelatin, mix boiling water and gelatin powder as usual for gelatin optics. Then, add as much sugar as you can dissolve in the mixture while it is still hot. You can make a slab, or mold it in a cylinder (Fig. 4.10). Remember to coat the inside of any container you use with oil for easier removal.
2. After the gelatin has set, remove it from the container and cover it with cold water for a few hours. Sugar will diffuse from the surface in contact with the water, resulting in a gradually changing sugar concentration and a gradually changing index of refraction, higher in the center and lower toward the edges. This causes a beam of light to curve gradually in the material, rather than follow a straight line.

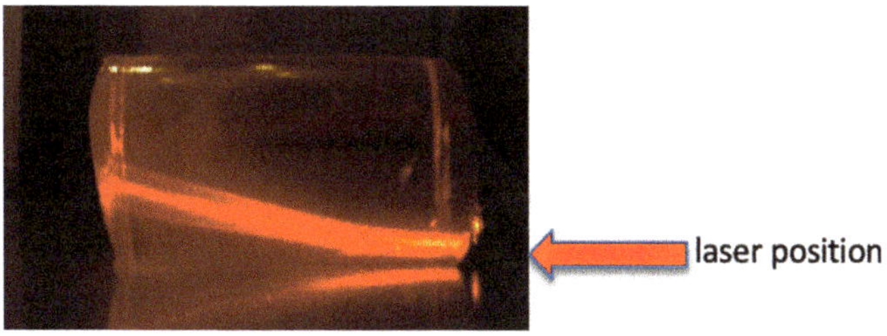

Figure 4.10 GRIN gelatin. This piece was molded in a large (4.5 cm diameter) plastic prescription medicine container. The laser beam enters horizontally from the lower right and curves upward as it travels to the left inside of the gelatin.

The Magic Lens

CAUTION! Do not look into the laser or at any reflections of the laser from shiny surfaces.

<u>Question:</u> Can a double convex lens (convex on both sides) make light diverge?

Physics books often include ray diagrams such as the one below showing how a beam of light behaves as it passes through a lens. The ray diagram might lead you to believe that a lens that is thicker in the middle will always cause light to converge to a focus point. Is this always true?

Materials

- Two watch glasses (borrow some from a chemistry laboratory if you can); suppliers are listed in the resources chapter
- Tube of silicone caulk, such as bathroom or aquarium caulk

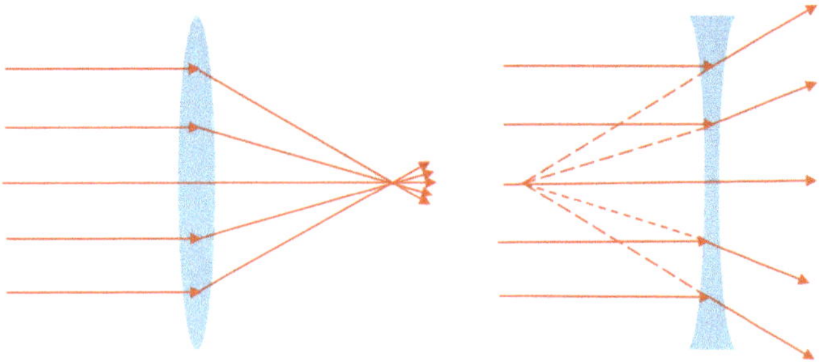

Figure 4.11 Several rays of light are shown passing through a cross section of glass lenses in air. *Left*: A converging lens brings rays to a focus point. *Right*: A diverging lens spreads rays as if they came from a point on the left side.

- Fish tank large enough to hold the lens submerged in water. A large plastic shoebox could also work. You want the lens to stand completely under water.
- Laser pointer (A ray box is even better!)
- Clamps to hold the lens in place in the tank (optional)

Science facts

A lens works by refraction, the bending of light as it goes from one material into another where it changes speed. The law of refraction tells us the direction that light bends when it goes from one material to another. The angles that a ray of light makes as it enters and exits the material are measured from a normal line, that is, a line that makes a 90-deg angle to the surface. If light travels more slowly after it crosses the surface, it bends toward the normal line. If it travels faster on the other side, it bends away from the normal line. This behavior explains how the lens can focus light to a point.

Figure 4.12 shows a beam of light from a laser entering a plate of glass. The laser light bends toward the normal line when it enters the glass and away from the normal line when it leaves the glass (not shown).

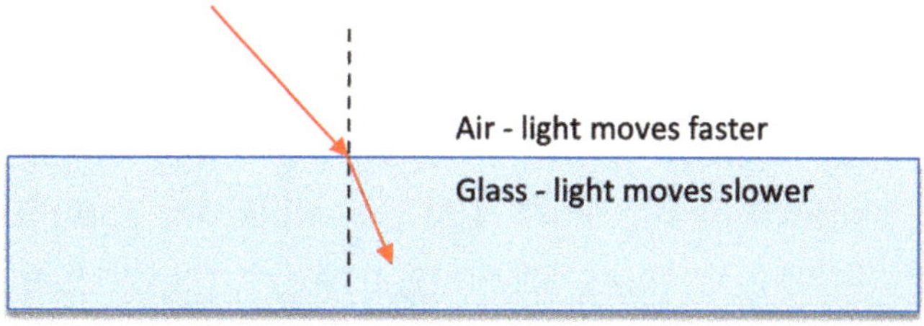

Figure 4.12 Light bends toward the normal (dashed) line when it enters a material where it travels more slowly (air to glass in this case). Which way will it bend when it enters a material where it travels faster (glass to air)?

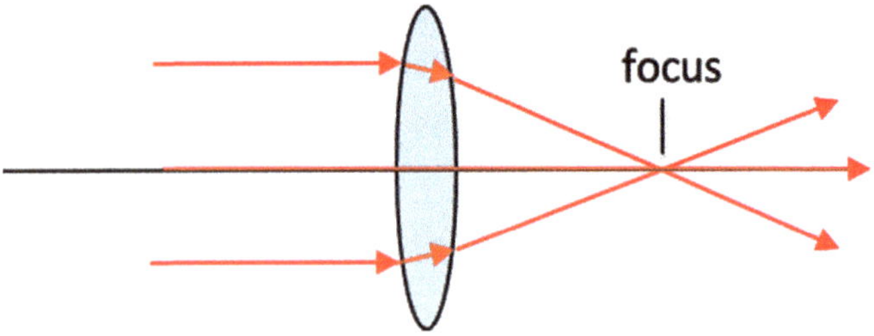

Figure 4.13 A converging lens brings light rays to a focus point on the other side.[4]

When used in air, a lens made of glass that is thicker in the center than at the edges will cause light rays to bend so that after passing through the lens they meet at a point on the other side. Light travels more slowly in glass than in air. We call this type of lens a converging lens. Figure 4.13 shows how the light bends as it enters the glass from the air and then leaves the glass to return to the air. The point where the light rays meet is called the focal point of the lens.

Procedure

1. First you need to make the "air lens." Clean the two watch glasses, then carefully squeeze a thick bead of caulk all the way around the edge of one of them. Be sure that there are no gaps or your lens will leak! Gently push the second watch glass onto the first so that their edges are sealed together as shown in Fig. 4.14. Let the caulk dry completely.
2. When the caulk is completely dry, fill the fish tank and add just a drop or two of milk to the water. This helps you see the laser light better.

3. What do you think will happen when light goes through the air lens when it is under water? (You'll need to hold the lens in place—it will float!) Shine a laser pointer into the lens at several places, keeping the beam straight and parallel to the table (Fig. 4.15). If you have a glass or plastic lens, repeat the experiment with that lens in the tank of water. (A large magnifier with a handle works well.)

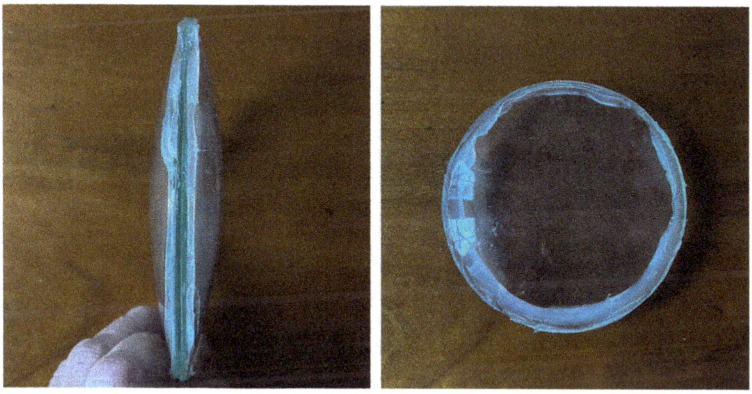

Figure 4.14 Construction of the "air lens." Be sure to seal the edges completely so water can't get inside!

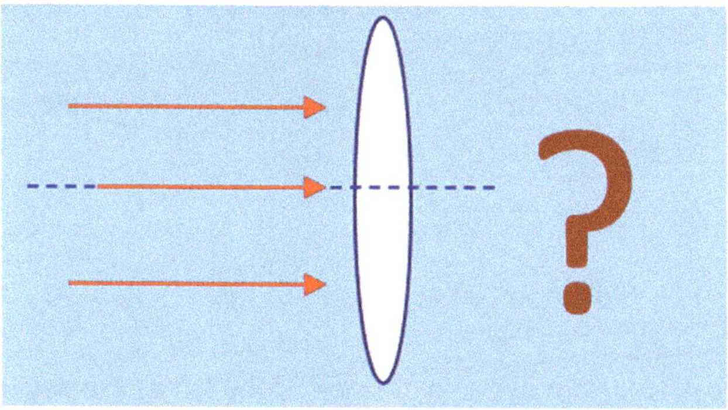

Figure 4.15 What happens when you shine laser beams through a lens made of air under water?[4]

Observations

Does the submerged air lens cause light to converge to a point or does it make the light spread out (diverge)? How would light behave with a solid glass lens under water?

Conclusion

Is it true that a lens that is thicker in the middle than at the edges *always* focuses light to a point? What else besides the shape of a lens determines how light bends when passing through it?

Sound travels faster in plastic than it does in water. What shape should a plastic lens have in order to focus an underwater beam of ultrasound?

Application—the misbehaving lens

How do swim goggles improve your vision under water? Your eyes focus light onto your retina, where sensors (rods and cones) detect the image and send the information on to your brain (Fig. 4.16). But most of the focusing is actually done by the cornea, rather than

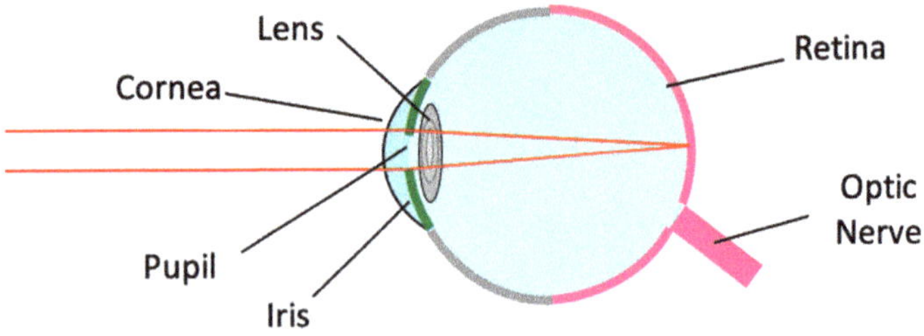

Figure 4.16 Normal eye in air. Most of the focusing power is due to the difference in index of refraction between the curved cornea and the surrounding air.

the lens, because lens power depends in large part on the difference in the index of refraction (speed of light) of the lens compared to that of the surrounding medium. The eye's lens is surrounded by fluids whose index of refraction is not that much different from the lens, but the cornea is normally surrounded by air.

When you open your eyes under water, your vision is blurry because the difference in index of refraction between water and your eye is not enough to focus light on the retina—you become severely hyperopic, or far-sighted (Fig. 4.17). Swim goggles restore the air film in front of your eye and allow the cornea to do its job (Fig. 4.18).

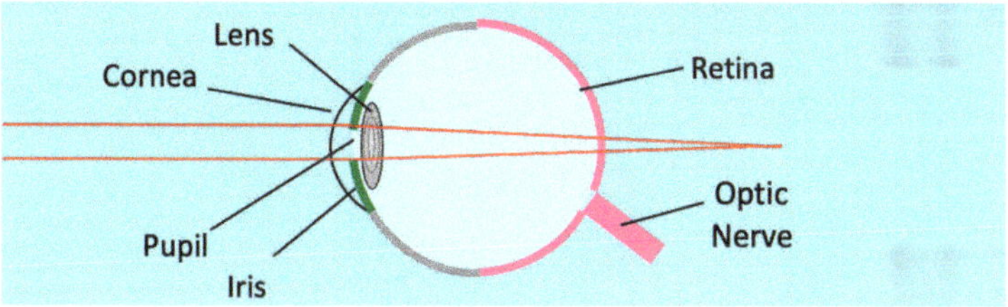

Figure 4.17 Under water, the index of refraction change between the water and the cornea is small, and light from a distant object focuses behind the retina. It's similar to being farsighted—everything looks blurry.

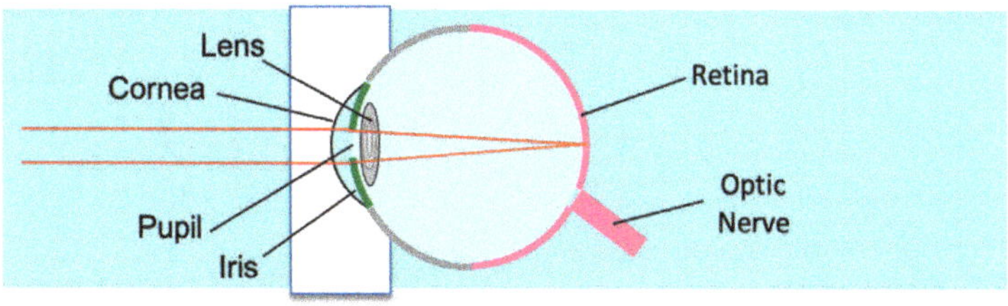

Figure 4.18 Swim goggles (white rectangle) restore the air film at the cornea, allowing the eye to focus.

Chapter 5
Exploring Diffraction

How Thick Is Your Hair?

CAUTION! Do not look into the laser or at any reflections of the laser from shiny surfaces.

Question: How can you use a laser to measure the thickness of your hair?

Materials

- Laser pointer
- Tape measure or ruler
- 2 clothes pins or a partner to hold the laser steady
- Paper and tape

Science facts

When waves encounter obstacles, they change direction. This is called *diffraction*, and it applies to all kinds of waves, including light, sound, electromagnetic, and even water waves. For example, ocean waves passing through a narrow gap will spread out on the other side, as seen in Fig. 5.1.

Figure 5.1 Ocean waves bend around an obstacle as they pass through a gap in a breakwater, forming circular waves on the other side (credit: Dmitris1 CC BY-SA 4.0 via Wikimedia Commons).

Diffraction of light is easiest to see with a laser, but you can see diffraction of ordinary light by holding two pencils close together side-by-side and looking at a light source through the tiny crack between them. A blank computer screen or blue sky works well as a light source, and you need to put your eye a few centimeters from the pencils. With practice, you will see dark and light bands that we call *fringes* (Fig. 5.2). The bands are caused by the diffraction of light as it passes through the tiny opening.

You can also see everyday diffraction when light from a streetlamp is viewed through the tiny holes in the fabric of sheer curtains. In this case, the different wavelengths

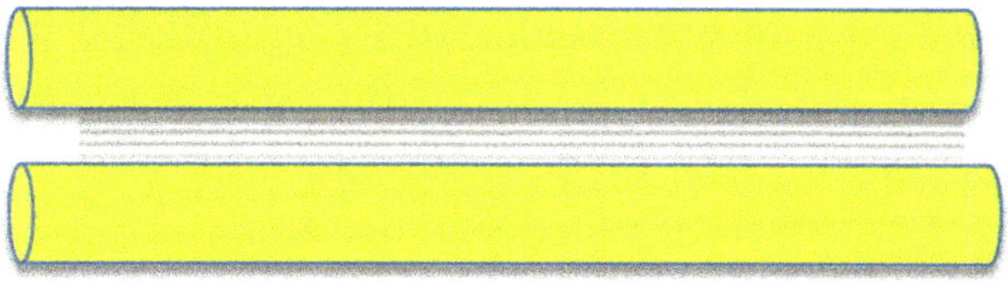

Figure 5.2 Representation of the fringes between two yellow pencils. This shows only a small part of the pencil, and the crack between them is only a millimeter or so across. The fringes are parallel to the crack.

of white light bend different amounts and interfere with neighboring waves, causing interesting patterns, as shown in Fig. 5.3.

Figure 5.3 A neighbor's security light viewed through sheer curtain fabric. Diffraction of white light passing through the tiny holes between threads of the fabric creates a colorful pattern. Can you see other examples of diffraction in this photo?

Scientists have developed equations that predict what the diffraction pattern will look like when light passes through or around certain specific obstacles. In this activity you will shine laser light around a single hair and observe a pattern of dark and bright spots. We'll measure the spacing of these spots and calculate the width (diameter) of the hair.

To do the calculation we need a way to tell which fringe, or dark spot, we're talking about. They all look about the same, so we number them starting at the center of the pattern. We'll be naming the dark spots between the bright red ones. The first dark fringe nearest the center of the pattern is $m = 1$, the next is $m = 2$, and so on. (The letter m is commonly used to indicate a counting number such as 1, 2, 3, and so on.) Figure 5.4 shows how the dark fringes are numbered. You will actually see many more fringes than in the drawing on the left in Fig. 5.4, and they will probably not be sharp lines. The center of the pattern will be a little blurry, as you can see in the photograph of an actual diffraction pattern made by a hair on the right of Fig. 5.4.

The formula to calculate the diameter of the hair is

$$d = \frac{m\lambda x}{y}$$

In this formula:

$d =$ diameter of the hair

$m =$ the order, or number from the center, of the dark fringe you measured (1, 2, 3, etc.)

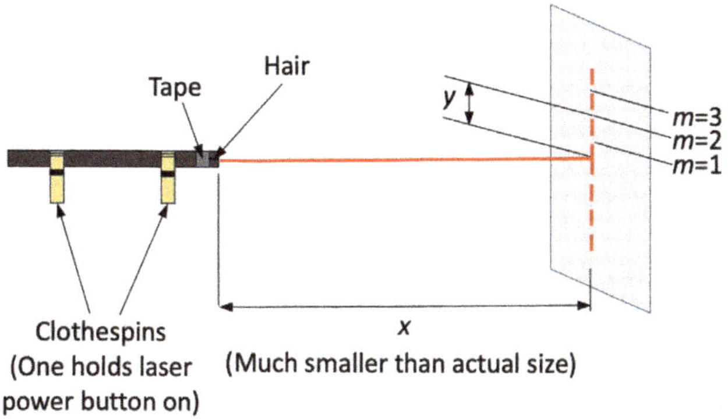

Figure 5.4 *Left*: Diffraction pattern caused by a hair. *x* is the distance from the hair taped across the laser aperture to the wall or screen. *y* in this diagram is the distance from the (bright) center of the pattern to the second, or $m = 2$, dark fringe. *Right*: Photograph of the hair diffraction pattern taken about 2 meters from the end of the laser.[4]

λ = the wavelength of the light (This is written on the laser warning label.)

x = the distance from the hair to the wall or screen

y = the distance from the bright center of the pattern to the dark fringe you are measuring

You can see that you need to know which fringe you are looking at (m) as well as the distance from the center of the pattern to the fringe.

Procedure and observations

1. Tape a piece of hair across the output aperture of a laser pointer. Be sure that the laser is off when you

do this! Support the laser pointer using two clothespins as shown in Fig. 5.4.

2. Point the laser pointer at a wall (approximately 2–3 meters away, far enough that the fringes are separate enough to measure). Tape a piece of paper to the wall to serve as a screen that you can write on. (Please don't write on the wall!)
3. Turn on the laser pointer and observe the pattern formed on the wall.

   What does this pattern look like? Does it look like the shadow of a hair, or is it more complex? Are there many fringes or just a few? Are they clear or blurry?
4. With a pencil, mark the center of the pattern. Call this $y = 0$. Mark the positions of several dark fringes on either side and number them ($m = 1$ for the closest to the center, then $m = 2$, etc.).
5. Measure the distance from the laser pointer to the wall and record the value of x. Now you can take the paper down.
6. Measure the distance between the center of the pattern ($y = 0$) and one of the dark fringes you marked. Also record m, the order of the fringe whose distance you measured. Record the values of y and m.
7. Record the wavelength of the laser. It should be listed on the side of the laser pointer (typically 650 nm). Be sure the laser is off when you look for the label!
8. Use the equation to calculate the diameter of the hair. It should be approximately 70–100 μm (one micrometer = 1/1,000,000 of a meter).

Conclusion

Compare your results with those of your classmates. Is there a difference between blonde and brown hair? Some students have measured the size of their pets' hair! What might be a practical use of this experiment?

Data for the "How Thick Is Your Hair?" experiment

Type of hair you used (color, coarse or fine, etc.):

____

$x =$ ____________________

$m =$ ____________________

$y =$ ____________________

$\lambda =$ ____________________

Calculated hair thickness (diameter):

____

Exploring Resolution

Question: How close together can words be on a sign such that you can still read it from a distance? How close can stars be and still be seen as separate objects? Does it depend on wavelength (color)?

Materials

- Patterns of red dots and blue dots, 1 mm across, spaced 1 mm apart. Draw them on a piece of paper or print Fig. 5.7 on white paper using a color printer.
- Meter stick.

Science facts

When light passes through a small opening, it spreads out or diffracts. The diffraction pattern that results from light passing through a small round hole is a central bright spot (called the Airy disk) surrounded by dimmer rings of light around the disk in bull's eye fashion (Fig. 5.5). The size of the center disk depends on the wavelength (color) of the light, the size of the hole, and how far the hole is from the viewing screen. It might surprise you to learn that light passing through a small hole spreads farther than light passing through a larger hole! Red light also spreads out wider than blue light.

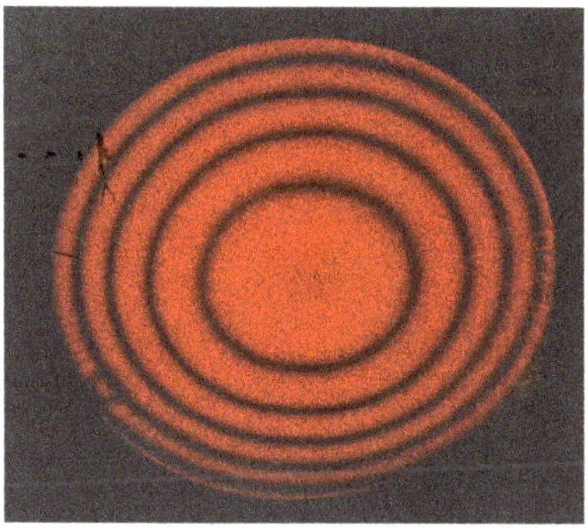

Figure 5.5 The diffraction pattern produced when red light from a helium-neon laser shines through a small (50 mm) pinhole and onto a piece of paper. The wide spot in the center is the Airy disk. (Photograph courtesy of Dr. Nicholas Massa.)

When light from two small points of light (for example, stars) passes through the same hole (such as the pupil of your eye), the light from each source will produce its own diffraction disk. *Rayleigh's criterion* states that you will be able to tell that there are two small dots rather than one dot when the center of one diffraction disk falls on the edge of the other. (Fig. 5.6) If the dots are closer than that or farther away, the two spots will blur together. Of course, this assumes that the system making the image of the dots (your eye) is otherwise perfect, which isn't the case. Rayleigh's criterion also applies to the lenses of instruments such as microscopes and telescopes. Even though they can have large apertures, tiny points of light at a large distance can still appear as one blur.

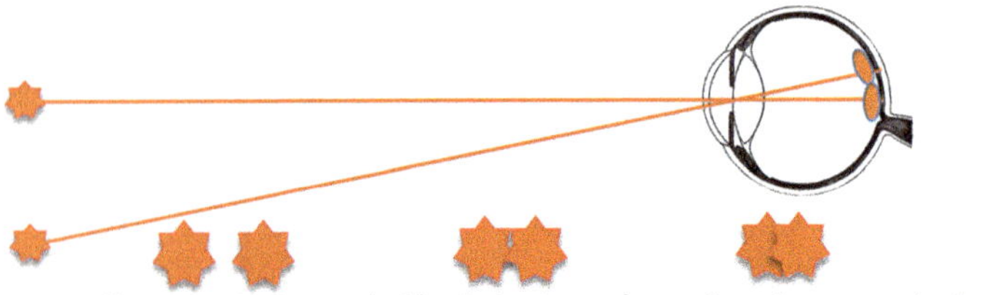

Figure 5.6 Rayleigh's criterion. When the diffraction disks are overlapping, the two stars appear as one blob. In this drawing, the size of the disks is greatly exaggerated and, of course, the stars are really much farther away.[4]

Procedure and observations

1. Copy or print the pattern shown in Fig. 5.7 in color. Tape the pattern on a wall.
2. Stand far enough away from the patterns in a well-lit room that they appear to be solid lines; that is, you can't see separate dots. Walk slowly toward the patterns. Carefully observe the patterns. At some point can you see separate dots instead of solid lines? Which dots can you see first, red or blue?
3. Record the distance where you can see the separate blue dots and the distance where you can see the separate red dots.

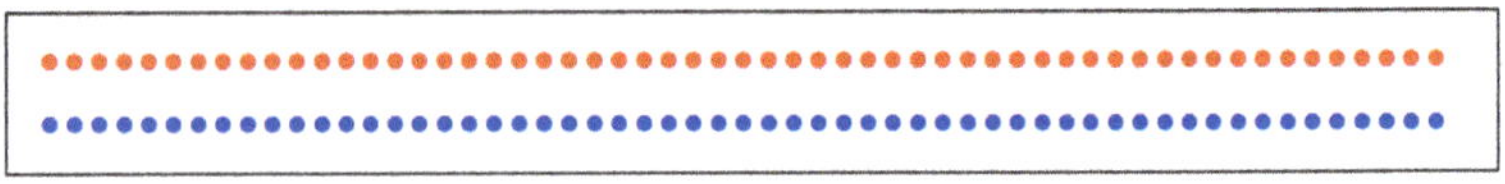

Figure 5.7 Red and blue dot pattern.[4]

Conclusion

Does the distance at which the spots can be resolved depend on color (wavelength)? If so, how?

Application

Diffraction and Rayleigh's criterion limit how small letters can be on highway signs and how close together lights on an airport tower can be. It also explains the style of painting called *Pointillism*, or *stippling*. In these paintings, millions of tiny dots merge together when viewed from a distance. In fact, two neighboring dots of different colors can appear to form a third color when viewed from far away. How can this be? If you live near an art museum, see if they have any pointillistic paintings on display and note at what distance the different color dots begin to merge.

What other very common devices depend on colored dots appearing to merge together?

Chapter 6
Polarization

Exploring Polarized Light

Questions: What is polarized light? How is polarized light produced? What is polarized light used for?

Materials

- You will need a polarizing filter to detect polarized light. Polarized sunglasses will do if you have them. Photographers often have polarizing filters, but they can be expensive. For some experiments you can use one polarizer and the light from an LCD screen such as a tablet or laptop screen. (This is noted in the polarizer instructions where appropriate.) See the Resources for other sources of polarizers, including so-called "3D glasses," which can be bought in bulk.
- Bowl of water (or a lake if you're near one). You can also use a polished wood or tile floor.
- Transparent plastic objects: ruler, protractor, comb, etc.

Science facts

Imagine that you and a friend are holding the ends of a rope. You shake your end up and down, and waves travel

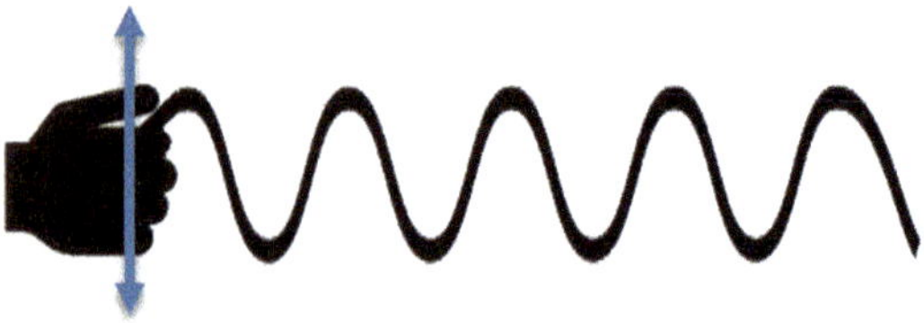

Figure 6.1 Shaking the end of a rope to make waves.

down the rope to your friend. You could also shake the end of the rope side to side or in any other direction, and the waves you make will change their vibration direction to match (Fig. 6.1).

Like the waves you made on a rope, the vibrations of light waves are perpendicular to the direction that the light is traveling. *Natural* or *randomly polarized* light has waves that vibrate in all directions, and the direction of vibration changes randomly over time. In Fig. 6.2, the arrows in the diagram on the left represent natural, randomly polarized light. The wave is traveling in the direction of the blue arrow, and some of the vibration directions are shown in red.

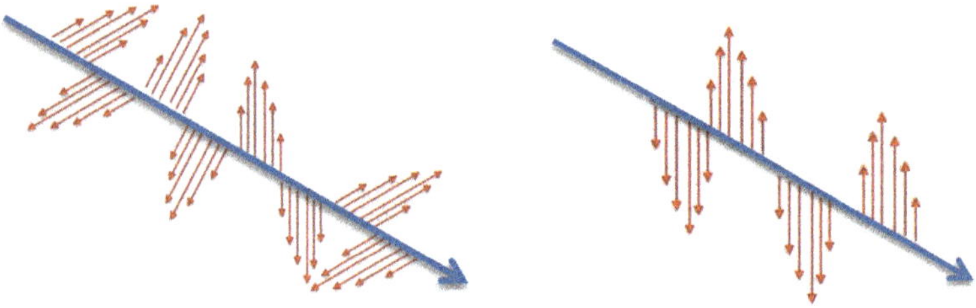

Figure 6.2 Natural (*left*) and vertically polarized (*right*) light. The blue arrows show the direction that light is traveling. The red arrows represent the direction of vibration of light waves—on the left, waves vibrate randomly in many directions, and on the right, all waves are vibrating in the same direction, in this case vertically.[4]

The diagram on the right represents *polarized* light, with only up and down vibrations. When light is polarized, the vibrations are in one direction only. You may wonder if polarized light is unusual. In fact, it's around you all the time! Polarized light may be produced by passing it through a polarizing filter (sometimes called a polarizer), scattering it from molecules (such as air molecules), reflecting it from a nonconductive surface, and/or passing light through certain crystals. In this activity you will make and observe polarized light.

Procedure and observations

Activity 1—Reflecting polarized light

1. Hold the polarizing filter (polarizer) in front of your eye and look at the still surface of a bowl of water (Fig. 6.3). (You can also look at a polished floor.) Rotate the polarizer while looking through it.

Observations

- What do you see? Does the reflection from the surface of the water change?
- When you can see the brightest reflection, the polarizer is allowing light wave vibrations to pass. The waves reflected from the water are vibrating mostly horizontally so they can pass through the filter when the *transmission axis* is horizontal. Like the slats in a fence, the polarizer is aligned to let this vibration through.

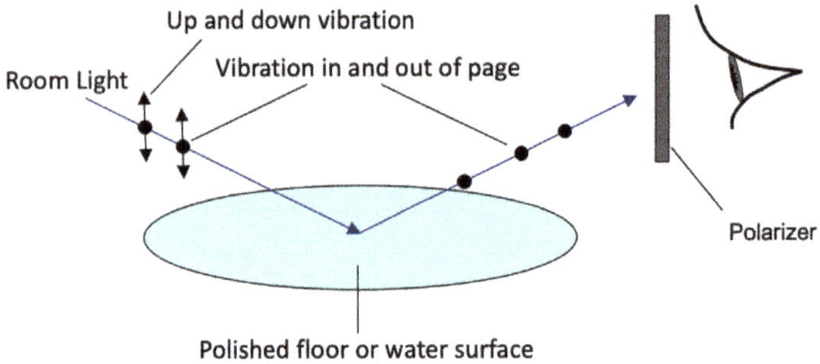

Figure 6.3 Reflecting polarized light. Randomly polarized light (coming from the left) is reflected so that it is polarized (on the right). The double-tailed arrows represent vibration in the plane of the page. The dots mean vibration in and out of the page (they represent the tail of an arrow). Reflected light from a surface such as water, ice, or even a polished floor mainly has wave vibrations in one direction. (The viewing angle determines how polarized the reflected light is. We'll ignore this.)[4]

2. Mark the polarizer with a piece of tape so you know the direction of the transmission axis (horizontal, when light is brightest). If you are using sunglasses, be careful not to damage the lens.
3. Repeat with the other polarizer. If you only have one polarizer, skip this step.

Activity 2—Polarizer pair

1. Now that you have labeled the transmission axis of each polarizer, look through one of them at a source of light (a lamp is fine). The light you see passing through the polarizer is linearly polarized; that is, the waves all vibrate in the same direction. You can change the direction of polarization by rotating the plastic, but your eye can't see the difference because humans can't sense polarization.

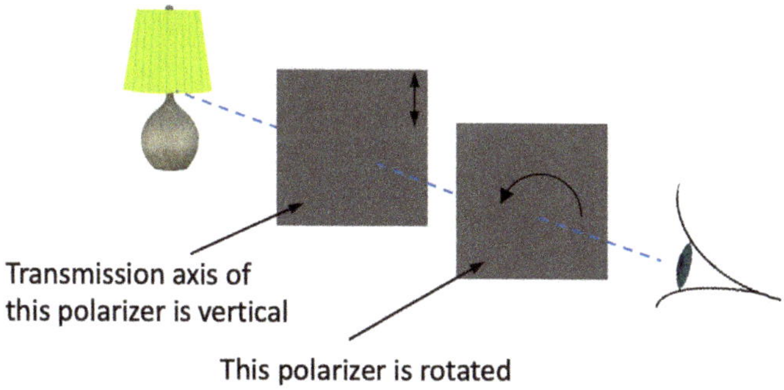

Figure 6.4 Looking through two polarizers. (You can also use one polarizer and an LCD screen.)[4]

2. Hold the polarizer so that it is producing vertically polarized light. (The transmission axis you marked with tape is vertical.) Place the second polarizer in front of the first, closer to your eye, with its transmission axis also vertical. Without moving the first polarizer, rotate the one closer to your eye through 360 deg (Fig. 6.4).

If you have only one polarizer (say, one piece of polarizing filter or a single pair of polarized glasses), you can rotate it in front of an LCD screen such as a laptop or TV screen. The light from a screen is polarized so you don't need the lamp or a second polarizer in this case.

Observations

- How does the lamp appear when you look through one polarizer, compared to looking directly at it without a polarizer?
- What happens when you rotate the front polarizer?
- How many times does the light dim during the 360-deg rotation?

Conclusion

How does the polarization of light explain your observations?

How do polarized sunglasses work to cut down glare? Does it matter in what direction the lenses are mounted in the frames?

If you see some really inexpensive glasses in a dollar store marked "polarized," how can you find out if they really are polarizing or just have a dark tint? Are UV-blocking sunglasses necessarily polarized?

Activity 3—Polarization by scattering

When light is scattered from very small particles such as molecules, the scattered light is polarized. This is called *Rayleigh scattering*. An easy-to-find molecular scatterer is air! On a sunny day, look at the blue sky through one of the polarizers or a pair of polarized sunglasses. Look at a patch of sky AWAY from the sun and rotate the polarizer in front of your eye as you look through it. Look at other parts of the sky as well. **DON'T LOOK AT THE SUN!**

Observation

- What do you see when you rotate the polarizer?

Conclusion

What does this observation tell you about the light scattered by the atmosphere? Bees and some other insects use the polarization of the sky to navigate (unlike human eyes, their eyes are adapted to sense polarization). They need to stay home if the sky is cloudy!

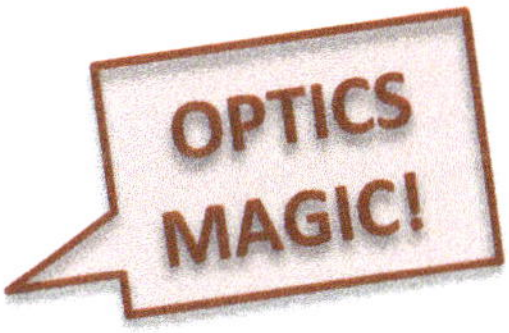

Activity 4—Stress patterns in plastic (changing the direction of polarization)

Once again, look through the two polarizers toward a lamp. Arrange them so the transmission axes are crossed (at right angles) and no light gets through. Place a piece of clear plastic such as a ruler or food wrap between the crossed polarizers, and twist or bend the plastic. Look through both polarizers and the film, like looking through a sandwich (Fig. 6.5).

You can also hold the plastic film in front of an LCD screen (such as a laptop or TV) and just use one polarizer. The screen provides polarized light and replaces the rear polarizer and the lamp.

Observations

- What do you see when you look through the crossed polarizers with nothing between them?
- What happens when the plastic is between the polarizers?

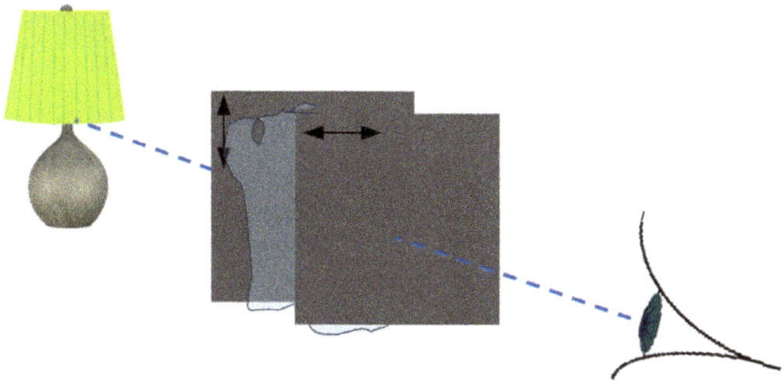

Figure 6.5 Looking through a "sandwich" of two polarizers with plastic in between.[4]

- Does what you see change if you bend or twist the plastic?
- What happens if you rotate one of the polarizers?

Conclusion

What do you think the plastic does to the polarized light passing through the first polarizer? Why do the colors appear?

Applications

Polarized light can be used to study stresses in a structure. If a model of a structure such as a bridge is built of transparent plastic, it can be weighted to simulate actual operating conditions. Viewing the model in polarized light helps to visualize where the structure will be stressed. Figure 6.6 shows some eyeglasses with plastic lenses placed between two polarizers with their transmission axes at 90 deg. You can see where the frame is squeezing the lenses.

Figure 6.6 Plastic lens eyeglasses viewed between two crossed polarizers.

Figure 6.7 A duck on a pond with (*top*) and without (*bottom*) a polarizing filter on the camera. (Photographs courtesy of Albert Yee, Philadelphia, Pennsylvania.)

Photographers also use polarizing filters. The two photos on the previous page were taken one after the other on a sunny day. For which one did the photographer place a polarizing filter over the camera lens? How can you tell? Why would a fisherman want to wear polarized sunglasses?

The Magic Box

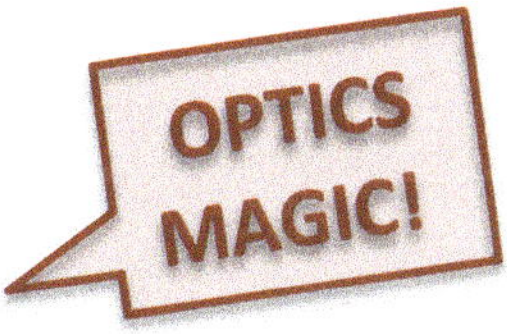

<u>Question</u>: Can you use light to make a magic box with a "wall that isn't there"? Where does the wall come from? Why can you push your finger through it with no resistance?

Materials

- A small rectangular cardboard box such as a shoe box or tissue box
- Four small (about 4 cm) square linearly polarizing filters (See Resources)
- Tape

Science facts

Before you do this activity, you should understand how polarized light works. You need to know the transmission axis of each of the four polarizers that you will use. Activity 1, Exploring Polarization, tells you how to do this.

Procedure and observations

1. Look through each polarizer, in turn, at the reflection from a shiny floor or water surface. Rotate the polarizer until the reflection is bright; the transmission axis will be horizontal. Mark the polarizers along the edge parallel to the transmission axis.
2. To build the magic box, cut rectangles from both the front and back sides of the cardboard box. Each rectangular opening should be just large enough to be completely covered by the two polarizer squares when they are placed side by side (Fig. 6.8). Carefully align these openings so you can look right through the box.

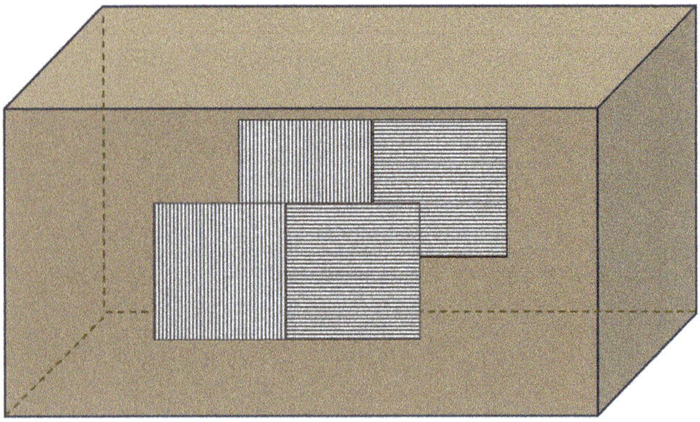

Figure 6.8 Construction of the magic box, showing the orientations of the four polarizers. Note that the transmission axes of matching pairs (left, right) are in the same direction on both the front and back of the box. The polarizers completely cover the holes cut in the box.[4]

Figure 6.9 Photograph of a “magic box” showing the apparent wall at the center where front and back polarizers overlap. This box has the polarizers taped to the inside of the box.

3. Tape two of the polarizing filter squares to the front opening. One filter should have its transmission axis in the vertical direction and the other in the horizontal direction. Look through the box and describe what you see.
   - Can you tell just by looking through them which polarizer is vertical and which is horizontal? Why or why not?
4. Tape the other two polarizing filter squares over the back opening. The orientation of the transmission axes is correct if, when viewed from the front, the vertical polarizers (front and back) are both on the same side (Fig. 6.8).
   - What do you see when you look through the box? When you look through both pairs of polarizers?
   - Does it matter if you look straight through (so the centers of the front and back of the box are aligned) or at an angle?
5. If one end of the box is open you can put your hand right "through" the wall, explaining to your audience that you have power over solid matter!

Conclusion

Where does the "wall" come from? Why do you need to use polarizers?

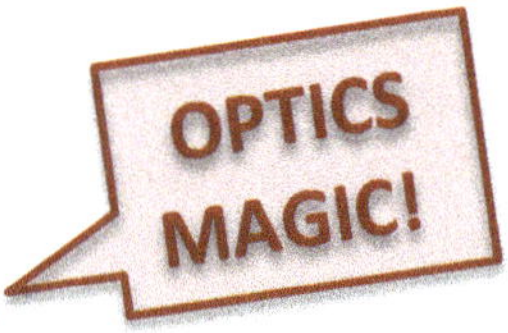

Polarized Light Art

<u>Question:</u> Can you use polarized light to create a colorful picture out of colorless plastic?

Materials

- Polarizing filter or sunglasses and an LCD monitor from a laptop or tablet. (You can also use two polarizing filters if you have them, but it really is easier with a laptop.)
- Pieces of cellophane tape (clear packing tape works well) or the cellophane wrapping material from produce such as lettuce or broccoli
- Optional: A small square of clear plastic the same size as the polarizers

Science facts

Before you use polarized light to create art, you should do the "Exploring Polarized Light" experiment on page 40. Here you learn that *natural* or randomly polarized light consists of light waves that vibrate in all directions. When light is polarized, the vibrations are in one direction only. In this exploration, you will create polarized light by using either a polarizing filter or the light from an LCD screen. The light from an LCD screen is polarized.

In "Exploring Polarized Light," you also learn that if you place two pieces of polarizing filter such that the transmission axes are parallel, nearly all of the light passes. If the transmission axes are at right angles, no light passes. But if you put certain types of material between the crossed polarizers, some light may get through and colors may appear from clear, colorless material.

In this activity we will use tape or cellophane to change the direction of polarization of light. (Some clear plastics also work—experiment!) You will make a "sandwich" with cellophane between an LCD screen and polarizer (or between two polarizers). Polarized light exits the screen, and then the direction of polarization is changed when it passes through the cellophane. The amount the polarization direction changes depends on the wavelength (color) of the light as well as the thickness and direction of the cellophane. If you then rotate the top polarizer, you will see different wavelengths at different angles of rotation (Fig. 6.10).

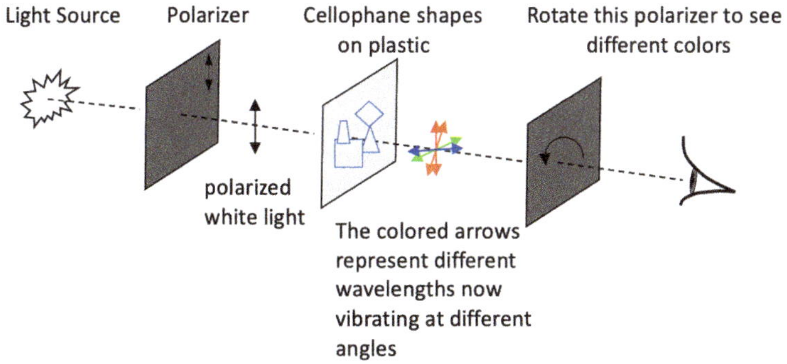

Figure 6.10 Colors are produced because the cellophane rotates the direction of polarization. How much the polarization is rotated depends on both the wavelength of the light and the thickness of the cellophane.[4]

Procedure

1. Cut the pieces of cellophane or tape into various shapes. Experiment by varying the thickness of the shapes from one to several layers.
2. Stick the tape onto the plastic square. If you use plain cellophane, you can attach it with a glue stick. The clear plastic is to protect the LCD screen from the sticky glue or tape. NOTE: Not everything sold as "cellophane" is really cellophane (which is made from cellulose). Some plastic films sold as cellophane don't work very well in this experiment. Try a variety of materials and see what works best!
3. When your piece of art is complete, place it against the LCD screen, preferably displaying a blank (white) document, or between two polarizing filters as shown in Fig. 6.10. Look at the piece you assembled through a polarizing filter or sunglasses. Rotate the filter—what do you see?

Conclusions

Did the thickness and direction of the cellophane make a difference in the color you saw? What other materials would this experiment work with besides cellophane?

Application

Austine Woods Comarow was a well-known artist who used this technique, which she called *Polage®*, to create beautiful works of art that change as you look at them through a moving polarizer. Some of her pieces cover complete walls in museums and other public spaces. You can see her art on her website austine.com.[13]

For more on polarization in nature, visit the website www.polarization.com.[14]

Chapter 7
Exploring Scattering

<u>Question:</u> **Why is the sky blue?**

Materials

- A fish tank is best, but we have done this successfully in a clear plastic shoe box about 33 cm long. The container should be glass or clear plastic.
- A flashlight. The kind you can focus to a narrow beam is best but not essential.
- Several drops of milk or cream.

Science facts

Why is the sky blue? Sunlight, as you know, contains all of the colors of the rainbow. When sunlight passes through the atmosphere, the shorter (bluer) wavelengths are scattered in all directions by tiny molecules of gas (mainly nitrogen and oxygen). Longer (redder) wavelengths travel in straighter lines with much less scattering. This is called *Rayleigh scattering*, which is also responsible for the polarization of the blue sky. "Sky blue" is not a single blue wavelength because the light from the sky includes a small amount of other scattered colors as well.

On a clear day, the sky directly above your head is deep blue because of the atmospheric scattering of blue light. But you may have noticed that, toward the horizon, the blue color of the sky becomes paler, or less *saturated.* This is because light travels through a much deeper layer of atmosphere to get to your eyes when you are looking toward the horizon (Fig. 7.1). Over this longer distance, the reds and greens have more time to scatter. The addition of the rest of the colors of the spectrum to the scattered blue light creates a whiter or paler shade of blue.

In this exploration, you will observe Rayleigh scattering in a fish tank filled with water to which a few drops of cream or milk have been added. An auditorium-sized version of this experiment starts at around 18 minutes in Professor Robert Greenler's scattering video.[15]

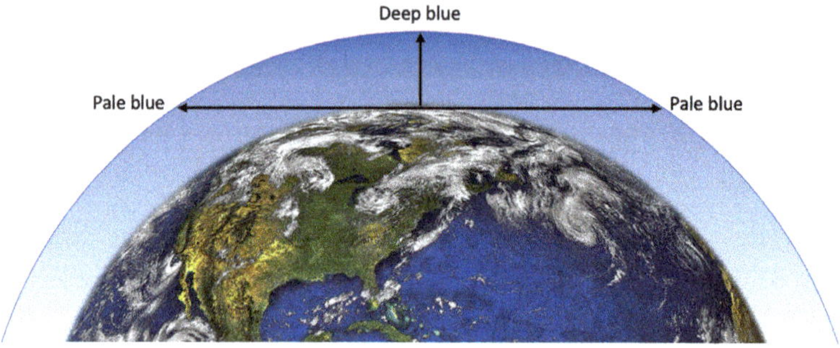

Figure 7.1 Light travels through a deeper layer of atmosphere when you look toward the horizon than when you look straight up (overhead). Note: Earth's atmosphere is much thinner than shown in this drawing, and it doesn't have a sharp edge.

Procedure

1. Fill the fish tank or other container with at least 5 cm of water (enough that the flashlight beam is fully under water) and add a few drops of milk. We found that 3 ml of milk per liter of water works well, but you should experiment to find the best ratio for your container. Mix the water and milk in the tank so that it looks just barely cloudy. Too much milk will ruin this experiment, so you need to add it slowly and observe the results before adding more. Note: You may be able to do this experiment in a tank with fish in it, especially if the water isn't crystal clear. Don't add milk and try to avoid the fish with the flashlight.
2. Shine the flashlight into one end of the tank as shown in Fig. 7.2.

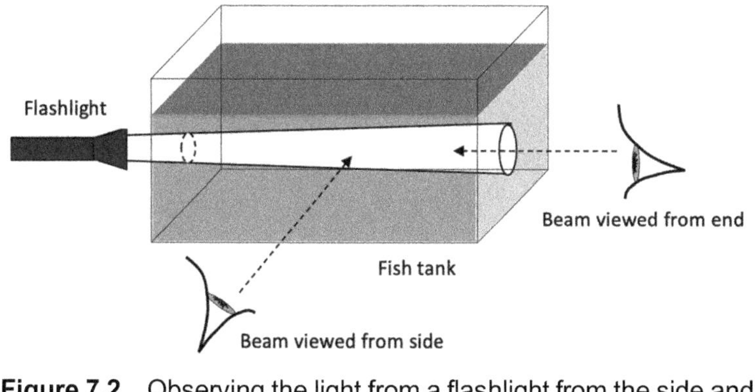

Figure 7.2 Observing the light from a flashlight from the side and the end of the tank of water.[4]

Observations

- Observe the color of the flashlight beam in the tank from the side of the tank. Do you see a color tint? What color? The effect will be very slight. It is helpful to dim the room lights to see the faint color.
- Look at the beam through the water from the end of the tank. What color is the light? Look briefly at the color of the flashlight bulb (not too long or you'll see spots for a while). Compare the color of the light when you look through the length of the tank to the color of the light when you look at it directly, not through the water. Explain the difference.

Conclusion

Which end of the light spectrum, blue or red, is scattered out of the beam in all directions in the tank? Which end of the spectrum, blue or red, goes with little scattering from one end of the tank to the other? In your own words, use the results of this experiment to explain why the sky is blue. Why do you think the sun appears red at sunset? (Hint: Look at Fig. 7.1.)

Research

If the shortest wavelengths are scattered the most, why isn't the sky violet? (Hint: Think about the spectrum of sunlight and how your eyes perceive color. To which colors are your eyes most sensitive? Least sensitive?)

Optional Exploration

Try this if you have already done the "Exploring Polarized Light" activity. Look through a polarizer at the beam from the side of the tank. Rotate the polarizing material in front of your eye (Fig. 7.3). What do you see? Look into the end of the tank, toward the light, and repeat. Where is the light polarized?

Is the blue sky polarized? Take your polarizer outside and look through it at the sky. DO NOT LOOK AT THE SUN! Some insects have special structures in their eyes that can detect polarization. How might being able to see polarized light help them navigate?

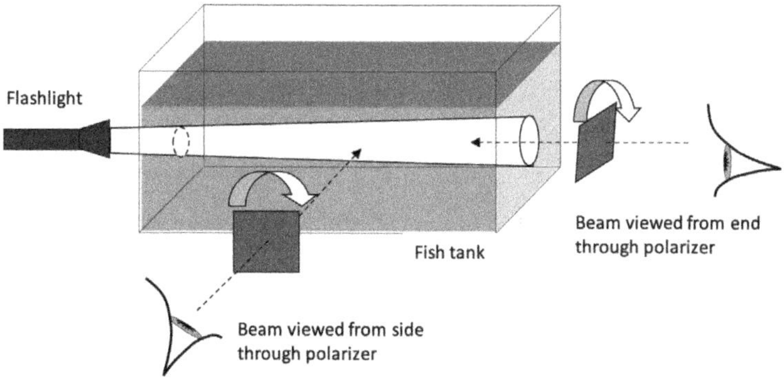

Figure 7.3 Observing the flashlight beam through a polarizer.[4]

Chapter 8
Exploring Laser Beams

CAUTION! Do not look into the laser cavity or at any reflections of the laser from shiny surfaces.

Question: Does laser light spread out as the beam travels through space? In what other ways is a laser beam different from a flashlight?

Materials

- Laser pointer
- Flashlight
- Diffraction grating or a CD or DVD
- Piece of waxed paper and a rubber band
- Meter stick and ruler

Science facts

Did you know that the word LASER is an acronym? It stands for **L**ight **A**mplification by **S**timulated **E**mission of **R**adiation. A laser is a special type of light source that has unique properties that make it very important for today's technologies. Unlike a flashlight, a laser typically produces only one wavelength (actually, an extremely small range of wavelengths). We say that lasers are *monochromatic* (one color). Laser light is extremely bright; it can be brighter than the sun! Laser light also spreads out very little when shined across a distance (see Fig. 8.1). Laser light is also *coherent*, which means that the waves in the beam stay "in step" with each other. This leads to interesting optical effects, as you will see.

The special properties of laser beams allow them to do the following:

- Travel very far distances (to the moon and back!)
- Be focused down to very small spots (to less than 1/10 of the size of a human hair) for cutting, welding, and drilling steel, ceramics, and other really hard materials

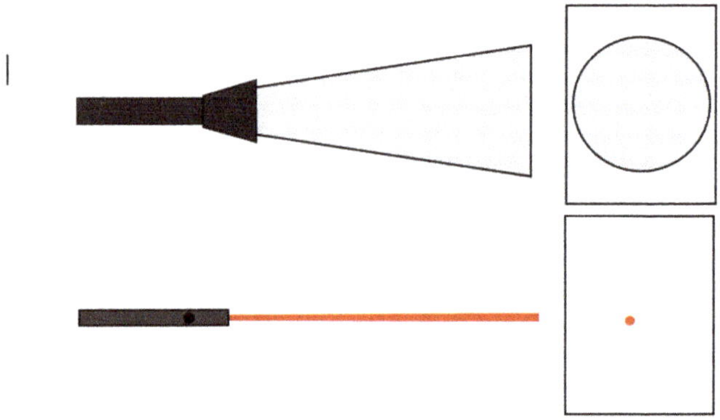

Figure 8.1 Flashlight versus laser pointer. Flashlights produce white light, while lasers pointers produce colored light. (Other types of lasers produce types of light you can't see, for example, UV or IR.) The spots of light produced by each type of light source may be seen by shining the light on a piece of paper taped to a wall.[4]

- Make very precise measurements (such as measuring the size of an atom)
- Cut tissue in laser surgery with minimal blood

You know that when you turn on the light bulb of a lamp in your home, the light spreads to illuminate the entire room. A flashlight beam, on the other hand, spreads less and illuminates only a small area. We call this spreading of a light *divergence*. Do you think a laser beam spreads out? After all, laser pointers are used to make small spots on a screen across the room!

In this exploration, you will find out if a laser beam diverges and learn some other ways that a laser is different from a flashlight.

Procedure and observations

1. First, you will look at laser divergence, or spreading. Working with a partner, hold the flashlight about one half meter from the wall and shine it onto a piece of paper taped to the wall. Mark the "edges" (diameter) of the beam. You will need to make your best guess about where the beam edges are! Repeat with the laser.
2. Move back as far as you can from the wall and shine the flashlight onto the piece of paper again. Again, mark the edges of the beam. Repeat with the laser. BE CAREFUL not to shine the laser in anyone's eye! Keep the beam well below eye level.

- Which beam spreads more, the flashlight or the laser? Did the laser spread at all? What would happen if you tried to shine the laser to hit a target 100 meters away? What do you think happened when scientists shined a laser at the moon?

3. Next, you will look at the color of laser light. If you have a diffraction grating, shine the flashlight beam through it and onto the piece of paper on the wall. Notice any colors that you see on the paper. Repeat with the laser pointer.
4. If you don't have a diffraction grating, reflect the flashlight beam from the CD or DVD onto the wall. What colors do you see? Repeat with the laser and carefully reflect the beam from the CD or DVD onto the paper on the wall. *Do not let the laser shine in anyone's eyes!*
   - What colors make up the flashlight beam? What colors make up the laser beam?
5. Finally, you will observe one of the effects of the coherence of laser light. Put the waxed paper over the end of the flashlight and fasten it with a rubber band. Shine it on the paper on the wall. Repeat with the laser.
   - What did the flashlight spot look like? Was it smooth or spotted? Did the laser light spot look different (aside from being red)? Do you notice any shimmering spots? (This is called laser *speckle.*)

Optional math activity

Use math to predict the size of a laser beam at a given distance from you.

You will need to carefully measure the diameter of the beam at two different distances from the lasers. You also need to measure the distance between the wall and the laser each time.

1. Hold the laser about one half meter from the wall so the beam is on the piece of paper. With your tape measure, measure the actual distance from the wall and record this distance as X_1 (see Fig. 8.2).
2. Measure the diameter of the beam. One way to do this is to mark what seem to be the edges of the beam on the paper with a pencil, then turn off the laser and measure the distance between the marks. Record this value as D_1.
3. Move back as far as you can from the wall and shine the laser onto the piece of paper again. BE CAREFUL not to shine the laser in anyone's eye! Keep the beam well below eye level.

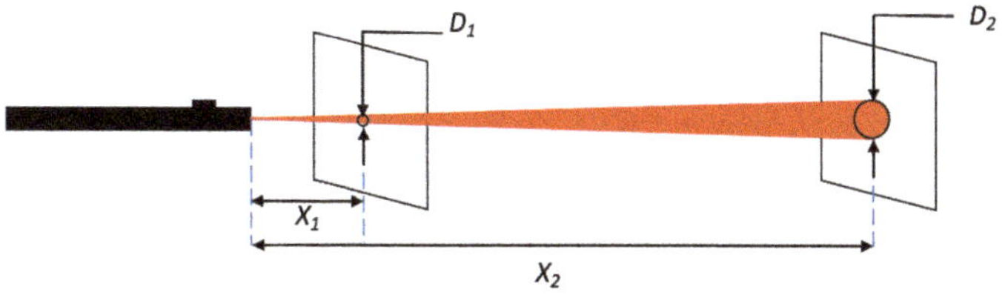

Figure 8.2 Measurements to determine the spreading of a laser beam.[4]

4. With your tape measure, measure the diameter of the beam again. Record this value as D_2.
5. With your tape measure, measure the distance from the wall. Record this value as X_2.

DATA from Math Activity

$X_1 =$ __________ $D_1 =$ __________

$X_2 =$ __________ $D_2 =$ __________

You can use the formula below to calculate how large the beam diameter will be at a different distance, X_3. In the formula, D_3 is the diameter of the beam at the distance X_3. Can you calculate what the beam diameter would be

1. 100 meters away?
2. 1609 meters away (1 mile)?
3. 385,000,000 meters away (239,000 miles away, the average distance to the moon)?

$$\text{Beam Diameter at } D_3 = \frac{(D_2 - D_1)}{(X_2 - X_1)} \bullet X_3$$

Conclusion

How is a laser beam different from a flashlight? Name as many differences as you can.

Why can't you shine a flashlight on the moon and have the reflection be received back on Earth? Note, even with a laser, it takes some clever measurement to actually detect the beam reflected from the moon. Scientists have been performing this experiment for more than 50 years.[16]

Chapter 9

Light You Can't See

Exploring Fluorescence and Phosphorescence

CAUTION! Do not stare at the UV light or shine it in someone's face.

<u>Questions:</u> What are fluorescence and phosphorescence? How are they similar? How are they different?

Materials

- Ultraviolet (UV) flashlight or UV LED (see the Resources chapter for where to find these)
- A red laser pointer
- Other sources of light such as colored LED finger lights (see the Resources chapter for sources)
- A square of glow-in-the-dark material (or glow-in-the-dark toys)
- Fluorescent minerals or other fluorescent materials. Many laundry detergents have whiteners that fluoresce. The ink that you are stamped with upon

entry to some venues is also fluorescent (see the Resources chapter for sources of this ink).

Science facts

The electrons in atoms are generally in their lowest energy state. Sometimes, though, they can be *excited* into a higher energy level by heat, a high voltage, absorbing light, or another means. Visible light can be produced when the excited electrons return to a lower energy level. The energy of the light *photon* is equal to the difference of the two energy levels, that is, to the amount of energy the electron loses when it goes to a lower level.

The excited electrons in a material may all give off their energy quickly—this is called *fluorescence*. For example, in a fluorescent light bulb, high-energy ultraviolet light produced inside the bulb causes the atoms in the bulb's coating to produce visible light. Sometimes the electrons release energy slowly, over a longer period of time. This process is known as *phosphorescence* and is what causes glow-in-the-dark materials to emit light for a long time.

The ultraviolet waves used in this experiment have a shorter wavelength and higher frequency than those of visible light. Frequency is the number of waves moving past a certain point in one second. Wavelength and frequency are related to the speed of light: the wavelength multiplied by the frequency equals the speed of light. So, as the wavelength gets shorter, the frequency increases. Higher frequency means greater energy, which is why you can get a sunburn from UV light but not from visible light.

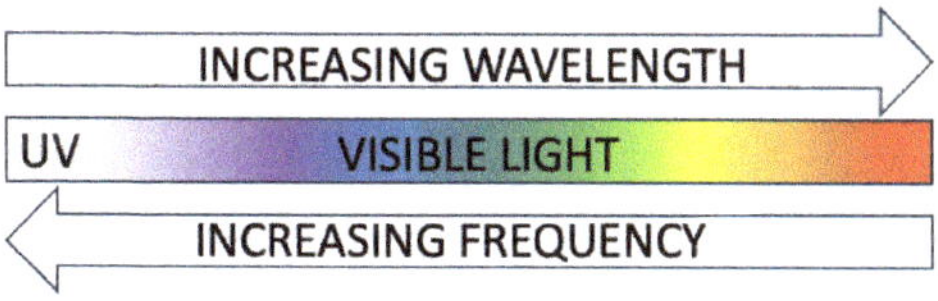

Figure 9.1 Visible and UV light spectrum showing the directions of increasing wavelength and frequency.

Part 1: Phosphorescence

Procedure

1. In a darkened room (it does not need to be completely dark), place the item with glow-in-the-dark material on a table. What do you think will happen if you shine each of the light sources—the laser, the UV light, a colored LED—on the material? Write down your predictions before you try it.
2. Test your predictions by shining each of the lights onto the material. Record:
   - Which color lights make the material glow. Why do only some colors make the material glow?
   - How long does the material glow after the light is removed?
   - Were there any surprises?

Part 2: Fluorescence

Procedure

1. Observe the fluorescent material under normal room lights. If you have whitening detergent, dab a bit on the back of your hand.

2. Now turn off the room lights so that the room is dark. Shine the UV light on the materials and observe and record any differences in appearances when the light is on compared to when it is off. Does the glow last after the light is removed?
3. Try the other colored lights—do you notice anything? Why does this only work with UV? (How is UV different from visible light?)

Conclusion

Both fluorescence and phosphorescence result from atoms releasing energy in the form of light. What is the difference between them? Why do you think detergent might be fluorescent?

Exploring UV Light and Sunscreen Lotion

CAUTION! Do not stare at the UV light or shine it in someone's face.

Question: How well does sunscreen protect against UV light?

Materials

- UV light (or you can use the sun—but do not look at the sun!)
- UV beads (See the Resources chapter for sources for UV color-change beads)
- Different types of sunscreen lotion, small pieces of glass or plastic wrap, and a brown plastic medicine container

Science facts

UV beads contain a pigment that responds to ultraviolet light by changing color. UV light has high energy and can cause skin damage such as sunburn, wrinkles, and skin cancer. Sunscreen lotions protect against UV by absorbing the light so it does not reach the skin.

Procedure

1. Expose the UV beads to the ultraviolet light or to the sun. What do you observe?
2. Return the beads to a dark place (or bring them in the house) until they turn white again. (Note that if UV beads are left in sunlight, they will eventually bleach and no longer change color.)

Or use a second group of beads. Place the beads under the glass or plastic and expose them to ultraviolet light or to the sun—do they still change color?

3. Spread the sunscreen on the glass or plastic wrap. (You can also put it directly on the beads, but it's messy!) Place the sunscreen-covered glass or plastic wrap over the beads and expose it to the UV light again. Be sure the beads are white before you begin this experiment. What do you observe? Is all of the UV light blocked? If you have different types of sunscreen lotion, compare the effects of different types.
4. Place a few of the white beads in the medicine container and expose the container to UV light. Do the beads change color? Why do you want to protect medications from UV light?

Conclusion

How does sunscreen protect your skin from ultraviolet light? Are all sunscreen lotions equally effective? Why is it important to protect your skin from exposure to ultraviolet light?

Chapter 10
Resources

Where to Find Supplies

Of course, the go-to place is often Amazon, but if you want to look locally, try a dollar store or big box store such as Walmart. Some items can be found in surprising places—such as laser pointers in pet shops.

1. Inexpensive **laser pointers** are available from a number of sources but be sure you are purchasing a *legal* pointer (<5 mW in the USA). One good source is your local pet store—they sell red laser pointers as "cat teasers." Dollar stores such as Dollar Tree often have laser pointers that are equipped with flashlights as well.
2. Color **LED finger lights**. Party stores and big box stores have these on their websites. If you want to buy in bulk, check sellers on www.amazon.com. Finger lights are the most convenient and inexpensive source of different color illumination.
3. **Ultraviolet (UV) flashlights**. Check Amazon for prices or visit your local sporting goods store.

4. Concave **watch glasses** can be found in chemistry labs. You certainly don't want to buy expensive ones, nor do you need them in bulk. If you can't find a chemist to give you a couple, try Amazon sellers or science supply stores such as Arbor Scientific (www.arborsci.com).
5. **Ray boxes** are very cool for optics experiments, but they can be expensive. You don't really need one for these activities; you can just use a couple of laser pointers (and a friend to help hold them). If you want to treat yourself to a ray box, Arbor Scientific has a set that includes a laser ray box and acrylic shapes that eliminate the need for gelatin.
6. You can buy **polarizers** at Educational Innovations (www.teachersource.com) if you don't want to use polarized sunglasses (which can be expensive). Rainbow Symphony (www.rainbowsymphony.com) sells polarizers mounted in slides and polarizing glasses for 3D movies (cut the cardboard frames in half to get two polarizers), but they only sell in quantity. Be sure the 3D glasses are linearly polarized, not circularly polarized. If you'd like a LOT of polarizing material, the www.polarization.com shop sells it by the foot. Their laminated polarizer is rigid and easy to handle.
7. You can make your own **diffraction grating** from a scrap of recordable CD. Scratch the metallic coating side and use tape to remove the coating—it pulls right off. With sturdy scissors you can cut it into smaller shapes if desired. If you really want to

buy a grating, Educational Innovations (www.teachersource.com) and Rainbow Symphony Store (www.rainbowsymphony.com) sell single-axis gratings mounted in slide frames and also in small sheets. (Single axis works best for experimenting.)

8. **UV color-change beads** are available at a number of vendors. We usually buy in quantity at Educational Innovations (www.teachersource.com). They come with an explanation of how they work.
9. **Fluorescent materials**. We like to use laundry detergent, the kind with whiteners (check the bottle), because it's easy to find. Olive oil is another common substance that changes color under UV light. Invisible (security) ink is available on Amazon and from other online vendors. Look for water-based ink to avoid sticky (but invisible!) spills.

References

For more comprehensive explanations of optics and procedures plus classroom-ready PowerPoint slides in English and Spanish, see *Dumpster Optics*.[3] There is no cost to download and use these lessons.

During the Covid19 pandemic in 2020, Optica hosted two series of webinars for PBL (Problem-Based Learning) Projects. They are linked from the PBL website.[5] The first five webinars listed were aimed mainly at high school teachers and college faculty participants of PBL Projects. The eight webinars in the lower list were based on the *Dumpster Optics* lessons and were offered to teachers, students, and home-schooling parents.

1. PBL Projects, YouTube, https://www.youtube.com/pblprojects
2. J. Donnelly, K. Amatrudo, K. Robinson, and F. Hanes, "The PHOTON explorations: sixteen activities, many uses," *Proc. SPIE* **9289**, 92892H, 12th Educational and Training in Optics and Photonics Conference (2014) [https://doi.org/10.1117/12.2070736].
3. "Dumpster Optics," *Problem Based Learning Projects*, www.pblprojects.org/dumpster-optics
4. *Explorations in Optics*, New England Board of Higher Education, Boston, Massachusetts, USA (2005).
5. "Resources from PBL/Optica Webinars," *PBL Projects*, pblprojects.org/resources-from-pbl-osa-webinars
6. "A comprehensive resource for safe and responsible laser use," Laser Pointer Safety, www.laserpointersafety.com
7. "Laser Toys: How to Keep Kids Safe," U.S. Food and Drug Administration (FDA), www.fda.gov/consumers/consumer-updates/laser-toys-how-keep-kids-safe
8. "Resources—Laser Safety Information—Laser Pointer Safety," The Laser Institute, www.lia.org/resources/laser-safety-information/laser-pointer-safety
9. "ChemCam on Mars—Curiosity Science Instruments," NASA, science.nasa.gov/mission/msl-curiosity/science-instruments
10. "Spectral Power Distribution," *Hyperphysics*, hyperphysics.phy-astr.gsu.edu/hbase/vision/spd.html
11. Stewart L.Woodruff, "Making Oatmeal Box Pinhole Cameras," *Oatmeal Box Pinhole Photography*, users.rcn.com/stewoody/makecam.htm
12. *The Brewster Kaleidoscope Society,* www.brewstersociety.com

13. *Austine Studios*, www.austine.com
14. *Polarization.com*, www.polarization.com
15. "Red Sunsets, Black Clouds and The Blue Moon: Light Scattering in the Atmosphere-UWM Science Bag," Vimeo, https://vimeo.com/139273980
16. "Laser Beams Reflected Between Earth and Moon Boost Science," NASA, www.nasa.gov/missions/laser-beams-reflected-between-earth-and-moon-boost-science/

Index

Judy Donnelly is retired from Three Rivers Community College (Norwich, CT) where she was a professor of physics and technology for 36 years. In 1997 she developed the Laser and Fiber Optic Technology (LFOT) associate degree program with the assistance of an industry advisory committee, and she served as LFOT Program Coordinator and optics instructor until retirement. She has been involved in STEM outreach since the mid 1980s when she ran an after-school STEM club, "Mrs. Wizard's Science Secrets," at her daughter's elementary school. She received the SPIE Maria Yzuel Educator Award (2003) and the Optica Esther Hoffman Beller Medal (2012) for outstanding contributions to optical science and engineering education. She is a Senior Member and Fellow Member of both societies. Since 1995, she has worked with the New England Board of Higher Education on projects funded by the National Science Foundation/Advanced Technological Education program to create curricula and provide faculty professional development for optics/photonics education and problem-based learning (PBL). She earned a B.S. degree from Tufts University and an M.S. from the University of Connecticut.

Nancy Magnani is currently an assistant science teacher for middle school students in the state of Washington. This includes developing curricula and facilitating the Science Club for middle school as well as providing optics and photonics demonstrations and lessons to both science club and photography club students. In addition, she has judged at local and state science fairs and volunteered at local museums in STEM education, including developing solar car challenges. While living in Connecticut, she was a Science Specialist/Grant

Facilitator for a regional education service center, provided science professional development to teachers, and wrote and facilitated STEM grant programs for 4th grade through high school students throughout eastern Connecticut. She presented optics and photonics workshops at National Science Teachers Association (NSTA), SPIE, and Optica (OSA) conferences. Partnering with Judy Donnelly, they presented optics and photonics workshops to students and adults throughout Connecticut. Prior to beginning a career in education, Nancy was a Senior Engineer-Transmission and Protection, designing and implementing high speed fiber optics systems for the telecommunications industry. Nancy is a senior member of both SPIE and Optica.